Mammalian Cell Biotec

The Practical Approach Series

Affinity Chromatography
Anaerobic Microbiology
Animal Cell Culture
Animal Virus Pathogenesis
Antibodies I and II
Biochemical Toxicology
Biological Membranes
Biosensors
Carbohydrate Analysis
Cell Growth and Division
Cellular Neurobiology
Cellular Calcium
Centrifugation (2nd Edition)
Clinical Immunology
Computers in Microbiology
Crystallization of Proteins and Nucleic Acids
Cytokines
Directed Mutagenesis
DNA Cloning I, II, and III
Drosophila
Electron Microscopy in Biology
Electron Microscopy in Molecular Biology
Essential Molecular Biology I and II
Fermentation
Flow Cytometry
Gel Electrophoresis of Nucleic Acids (2nd Edition)
Gel Electrophoresis of Proteins (2nd Edition)
Genome Analysis
HPLC of Small Molecules
HPLC of Macromolecules
Human Cytogenetics
Human Genetic Diseases
Immobilised Cells and Enzymes
Iodinated Density Gradient Media
Light Microscopy in Biology
Liposomes
Lymphocytes
Lymphokines and Interferons

Mammalian Development
Mammalian Cell Biotechnology
Medical Bacteriology
Medical Mycology
Microcomputers in Biology
Microcomputers in Physiology
Mitochondria
Molecular Neurobiology
Molecular Plant Pathology I and II
Mutagenicity Testing
Neurochemistry
Nucleic Acid and Protein Sequence Analysis
Nucleic Acids Hybridisation
Nucleic Acids Sequencing
Oligonucleotide Synthesis
PCR
Peptide Hormone Action
Peptide Hormone Secretion
Photosynthesis: Energy Transduction
Plant Cell Culture
Plant Molecular Biology
Plasmids
Post-implantation Mammalian Embryos
Prostaglandins and Related Substances
Protein Architecture
Protein Function
Protein Purification Applications
Protein Purification Methods
Protein Sequencing
Protein Structure
Proteolytic Enzymes
Radioisotopes in Biology
Receptor Biochemistry
Receptor–Effector Coupling
Ribosomes and Protein Synthesis
Solid Phase Peptide Synthesis
Spectrophotometry and Spectrofluorimetry
Steroid Hormones
Teratocarcinomas and Embryonic Stem Cells
Transcription and Translation
Virology
Yeast

Mammalian Cell Biotechnology

A Practical Approach

Edited by

M. BUTLER

Department of Microbiology,
University of Manitoba,
Winnipeg, Manitoba,
Canada

IRL PRESS
at
OXFORD UNIVERSITY PRESS
Oxford New York Tokyo

Oxford University Press, Walton Street, Oxford OX2 6DP

Oxford New York Toronto
Delhi Bombay Calcutta Madras Karachi
Petaling Jaya Singapore Hong Kong Tokyo
Nairobi Dar es Salaam Cape Town
Melbourne Auckland

and associated companies in
Berlin Ibadan

Oxford is a trade mark of Oxford University Press

Published in the United States
by Oxford University Press, New York

A catalogue record for this book is available from the British Library

Library of Congress Cataloging-in-Publication Data
Mammalian cell biotechnology : a practical approach / edited by M. Butler.
(The Practical approach series)
1. Animal cell biotechnology. 2. Mammals. I. Butler, M.
II. Series.
TP248.27.A53m36 1991 660'.6—dc20 90-27002

ISBN 0-19-963207-3
ISBN 0-19-963209-X (pbk)

Typeset by Footnote Graphics, Warminster, Wilts
Printed in Great Britain by
Information Press, Eynsham, Oxon

Preface

The development of a technology for the *in vitro* culture of animal cells has arisen from the increased demand for the commercially valuable products that are now available. Vaccines, monoclonal antibodies, lymphokines, and therapeutic enzymes are probably the best known of such products (biologicals). The process strategies that have allowed the large-scale production of these biologicals have arisen from a marriage of traditional academic disciplines. The biochemist, microbiologist, and chemical engineer have all come together in a unique exercise of communication to develop such strategies. This has allowed some small-scale laboratory techniques to aspire to large-scale industrial processes.

This book is intended to demonstrate the practical approaches of these diverse disciplines as they relate to the development of mammalian cell biotechnology. The diverse academic background of the authors is typical of those who have worked together to produce this development. Most of the authors have been members of or associated with the European Society of Animal Cell Technology (ESACT) or of the British Animal Cell Technology Club (SERC). Such groups have been instrumental in fostering the communication necessary for the development of the technology.

The book should be of value to the growing army of research workers who are now becoming involved in mammalian cell biotechnology. It should suit students who are in the final stages of a degree/HND course or higher degree students involved in laboratory work in this field. The book should also be of benefit to those working in industry whose objectives include the development or improvement of strategies for the commercial production of biologicals from mammalian cells.

Manchester
August 1990

MICHAEL BUTLER

Contents

Contributors

M. BUTLER
Department of Microbiology, University of Manitoba, Winnipeg, Manitoba R3T 2N2, Canada.

A. J. DICKSON
Department of Biochemistry and Molecular Biology, University of Manchester, Manchester M13 9PL, UK.

A. FLETCHER
New South Wales Red Cross Blood Transfusion Service, 153 Clarence Street, Sydney, Australia.

B. GRIFFITHS
Division of Biologics, PHLS CAMR, Porton Down, Salisbury, Wiltshire SP4 0JG, UK.

A. HANDA-CORRIGAN
Department of Microbiology, University of Surrey, Guildford GUI 5XH, UK.

C. HARBOUR
Department of Infectious Diseases, University of Sydney, New South Wales 2006, Australia.

D. G. KILBURN
Department of Microbiology, University of British Columbia, Vancouver V6T 1W5, Canada.

N. JENKINS
Biological Laboratory, University of Kent, Canterbury, Kent CT2 7NJ, UK.

A. LYDDIATT
Department of Chemical Engineering, University of Birmingham, Edgbaston, Birmingham B15 2TT, UK.

C. MACDONALD
Department of Bioscience and Biotechnology, University of Strathclyde, Glasgow, G4 0NR, UK.

A. J. MACLEOD
Protein Fractionation Centre, Scottish Blood Transfusion Service, Edinburgh EH17 7QT, UK.

Abbreviations

AAV	adeno-associated virus
ABTS	2,2′-azino-di-(3-ethylbenzathiazoline) sulphonate
AcNPV	autographa californica nuclear polyhedrosis virus
Ad	adenovirus
ACTH	adenocorticotrophic hormone
ADH	anti-diuretic hormone (vasopressin)
APRT	adenine phosphoribosyl transferase
ATCC	American Type Culture Collection
BCPIP	5-bromo-4-chloro-3-indolylphosphate *p*-toluidine
BHK	baby hamster kidney
BME	basal medium (Eagle's)
BPV	bovine papillomavirus
BSA	bovine serum albumin
BUdR	5-bromodeoxyuridine
BSF	B cell stimulation factor
CAD	carbamoyl phosphate synthetase, aspartate transcarbamoylase, dihydroorotase
CAMR	Centre for Applied Microbiology Research (UK)
CAT	chloramphenicol acetyltransferase
CEA	carcinoembryonic antigen
CHO	Chinese hamster ovary
CMV	cytomegalovirus
CNS	central nervous system
Con A	Concanavalin A
CPG	controlled pore glass
CSF	colony stimulating factor
DCR	dominant acting control region
DEAE	diethylaminoethyl
DEPC	diethyl pyrocarbonate
DHFR	dihydrofolate reductase
DMAP	dimethyl aminopropyl
DMEM	Dulbecco's modification of Eagle's medium
DMSO	dimethylsulphoxide
DOT	dissolved oxygen tension
DOTMA	N-[1-(2,3 dioleyloxy)propyl]-N,N,N-trimethylammonium chloride
EBNA	EBV nuclear antigen
EBV	Epstein–Barr virus
ECACC	European Collection of Animal Cell Cultures

EC	extracapillary (space)
ECGF	endothelial cell growth factor
ECM	extracellular matrix
EDTA	ethylenediamine tetraacetic acid
EGF	epidermal growth factor
ELISA	enzyme-linked immunosorbent assay
EPO	erythropoietin
eRDF	enriched RPMI/DMEM and F12 (medium)
FCA	Freund's complete adjuvant
FCS	fetal calf serum
FGF	fibroblast growth factor
FMDV	foot-and-mouth disease virus
FRhK6	fetal rhesus monkey kidney (cells)
FSH	follicle-stimulating hormone
G-CSF	granulocyte colony stimulating factor
GM-CSF	granulocyte–macrophage colony stimulating factor
GMEM	Glasgow's modification of Eagle's medium
GPI	glucose phosphate isomerase
GPK	guinea-pig keratocytes
GS	glutamine synthetase
HA	haemagglutinin
HAT	hypoxanthine, aminopterin, thymidine
HBS	Hepes buffered saline
HBsAG	hepatitis B surface antigen
HDC	human diploid cells
HeLa	Henrietta Lach (cells)
HEP	hepatoma
Hepes	4-(2-hydroxyethyl)-1-piperazineethane sulphonic acid
hGH	human growth hormone
HIV	human immunodeficiency virus
HLA	human leucocyte antigen
HGPRT	hypoxanthine–guanine phosphoribosyl transferase
HITES	hydrocortisone, insulin, transferrin, ethanolamine, and selenite
HSV	herpes simplex virus
HT	hypoxanthine, thymidine
IC	intracapillary (space)
IFN	interferon
Ig	immunoglobulin
IGF	insulin-like growth factor
IL	interleukin
IMDM	Iscove's modified Dulbecco's medium
IMP	inosine monophosphate
ITES	insulin, transferrin, ethanolamine, and selenite
JE(B)	Japanese encephalitis (B) virus

LAF	lymphocyte activating factor
LDH	lactate dehydrogenase
LH	luteinizing hormone
LTR	long terminal repeat
mAb	monoclonal antibody
M-CSF	macrophage colony stimulating factor
MDCK	Madin Darby canine kidney
MEM	minimal essential medium
m.o.i.	multiplicity of infection
MKC	monkey kidney cells
MoMLV	Moloney murine leukaemia virus
MRC	Medical Research Council (UK)
MS	mass spectrometer
MSA	multiplication-stimulating activity
MSX	methionine sulphoxamine
MTX	methotrexate
mol. wt	molecular weight
NBT	Nitroblue tetrazolium
NGF	nerve growth factor
NIH	National Institute of Health (USA)
NPC	nasal pharyngeal carcinoma
OTR	oxygen transfer rate
OUR	oxygen uptake rate
PAGE	polyacrylamide gel electrophoresis
PALA	N-(phosphoacetyl)-L-aspartate
PBS	phosphate buffered saline
PC	personal computer
PDGF	platelet-derived growth factor
PDL	population doubling
PEG	polyethylene glycol
PGD	phosphogluconate dehydrogenase
PHA	phytohaemagglutinin
PHLS	Public Health Laboratory Service
PID	proportional, integral, and differential (control)
PMA	phenyl mercuric acetate
PMSF	phenyl methyl sulphonyl fluoride
Rh	rhesus
RPMI	Roswell Park Memorial Institute (1640 medium)
RSV	Rous sarcoma virus
RT	reverse transcriptase
SDS	sodium dodecyl sulphate
SERC	Science and Engineering Research Council (UK)
SV	simian virus
TBE	tickborne encephalitis

TBS	Tris buffered saline
TCA	trichloroacetic acid
TCGF	T cell growth factor
TE	Tris, EDTA
TEMED	N,N,N′,N′-tetramethylethylenediamine
TGE	Tris, glucose, EDTA
TGF	transforming growth factor
TLCK	N-tosyl-L-lysyl chloromethane hydrochloride
TNF	tumour necrosis factor
TK	thymidine kinase
tPA	tissue-type plasminogen activator
TRF	T cell replacing factor
TSH	thyroid stimulating hormone (thyrotrophin)
UMP	uridine monophosphate
UR	uptake rate
WCB	working cell bank
WI	Wistar Institute

1

The characteristics and growth of cultured cells

MICHAEL BUTLER

1. Introduction

In recent years, mammalian cells have become the preferred cell type for the large-scale production of a range of commercially valuable compounds (biologicals). The technology developed to exploit animal cell production is closely associated with traditional microbial fermentation processes used for bacteria and fungi. The characteristics of animal cell cultures are, however, significantly different from such microbial cultures. In general, growth is slower, the cells are more fragile, and the nutrient requirements are more complex.

Animal cells have been studied at the laboratory scale for many years and several good texts are available to describe the practical techniques used (1–11). In this chapter, some of the background of animal cell technology will be explored and some of the characteristics of the most widely used cultures and processes will be highlighted.

2. The development of animal cell technology

The development of animal cell culture can be traced back to the work of Ross Harrison in 1907 on cell entrapment and growth from explants of frog embryo tissue (12). The cell growth that he observed in clotted lymph fluid in a depression slide is often regarded as the foundation of animal cell culture as a science (see *Table 1*). In his technique ('the hanging drop') the isolated tissue was suspended on the underside of a coverslip which was sealed over a depression in a microscope slide. Burrows continued the development of this technique with the use of plasma clots which were found to be more efficient for the growth of cells from warm-blooded animals. The matrix of insoluble protein provided the necessary anchor for cell growth, and nutrients were provided by the enclosed fluid.

One of the major difficulties of this work was the maintenance of the cultures free from contamination. The difference in growth rates between animal and bacterial cells is such that a low-level contamination in an animal

Table 1. Historical events in the development of cell culture

1885:	Roux maintained embryonic chick cells in a saline culture
1897:	Loeb demonstrated the survival of cells isolated from blood and connective tissue in serum and plasma.
1903:	Jolly observed cell division of salamander leucocytes *in vitro*
1907:	Harrison cultivated frog nerve cells in a lymph clot held by the 'hanging drop' method. Burrows improved this technique by the use of plasma clots
1913:	Carrel introduced strict aseptic techniques so that cells could be cultured for long periods
1916:	Rous and Jones introduced trypsin for the subculture of adherent cells
1923:	The 'Carrel' flask was introduced as the first specifically designed cell culture vessel
1940s:	The use of the antibiotics penicillin and streptomycin in culture medium decreased the problem of contamination in cell culture
1948:	Earle isolated mouse L fibroblasts which formed clones from single cells
1949:	Enders reported that polio virus could be grown on human embryonic cells in culture
1952:	Gey established a continuous cell line from a human cervical carcinoma known as HeLa
1955:	Eagle studied the nutrient requirements of selected cells in culture and established the first widely used chemically defined medium
1961:	Hayflick and Moorhead isolated human fibroblasts (WI-38) and showed that they have a finite lifespan in culture
1964:	Littlefield introduced the HAT medium for cell selection
1965:	Ham introduced the first serum-free medium which was able to support the growth of some cells
1965:	Harris and Watkins were able to fuse human and mice cells by the use of a virus
1975:	Kohler and Milstein produced the first hybridoma capable of secreting a monoclonal antibody
1978:	Sato established the basis for the development of serum-free media from cocktails of hormones and growth factors
1982:	Human insulin became the first recombinant protein to be licensed as a therapeutic agent
1985:	Human growth hormone produced from recombinant bacteria was accepted for therapeutic use
1987:	Tissue-type plasminogen activator (tPA) from recombinant animal cells became commercially available.

cell culture can quickly lead to bacterial overgrowth. This led Alexis Carrel, who was trained as a surgeon, to apply strict aseptic techniques to cell culture *in vitro*. He introduced the 'Carrel flask' which facilitated subculture under aseptic conditions and became the forerunner of modern tissue culture flasks. However, the procedures were elaborate and difficult to repeat and so cell culture was not adopted as a routine laboratory technique until much later.

In 1912 Carrel initiated a culture of chick embryo heart cells which were passaged for a reported period of 34 years. This led to the erroneous belief that, given the appropriate conditions, isolated cells could be cultured indefinitely. Later, analysis of Carrel's work showed that, as cell growth was maintained by the use of embryo extracts, new cells were being continuously added to the cultures during medium replenishment. The finite capacity for growth of 'normal' cells was eventually established from the work of Hayflick and Moorhead in 1961 (13).

A significant advance in the ability to initiate cultures was the use of trypsin to free cells from a tissue matrix and its later use for the subculture of anchorage-dependent cells. This technique, initially introduced by Rous and Jones in 1916, was not fully developed until the 1950s when its use enabled the establishment of homogeneous cell populations as opposed to tissue cultures which contain a mixture of cell types (14).

After the discovery in the 1940s of antibiotics such as penicillin and streptomycin their incorporation into the cell growth medium became an aid to the routine maintenance of asepsis. The biological fluids and embryo extracts used for cell growth were particularly vulnerable to contamination but supplementation with antibiotics reduced this problem. This encouraged the more widespread use of cell culture as a laboratory technique, particularly after the isolation of a variety of cell types which showed good growth characteristics *in vitro*. These included the chemically transformed mouse L cells and the human carcinoma cell line, HeLa.

These cell lines were the focus of Earle and Eagle's work in the 1950s on the development of chemically defined media (15, 16). The nutrient formulations developed replaced the undefined biological extracts previously used. This had the advantages of consistency between batches, ease of sterilization, and reduced chance of contamination.

The impetus for the application of the techniques of cell culture on a laboratory scale to large industrial-scale processes came with the capability of virus propagation on cell culture. In particular, the report by Enders *et al.* (17) that the poliomyelitis virus could be grown in cultures of human embryonic cells established the basis for the development of vaccine production and can be considered to mark the beginning of animal cell technology. The polio vaccine which was produced for mass vaccination in the 1950s became one of the first major commercial products of cultured animal cells. Since the 1950s a range of other products synthesized from animal cells have found commercial application. Consequently, the study of the optimization of culture conditions to maintain consistently high productivities from such animal cells *in vitro* has become of increased importance.

3. Characteristics of cell cultures

3.1 Primary cultures

A primary cell culture is established by inoculating growth medium with cells taken directly from animal tissue. The excised tissue is fragmented into small pieces with a forceps and scissors before placing into a sterile medium in a Petri dish. Treatment of the excised fragments with a proteolytic enzyme such as trypsin disaggregates the tissue into individual cells. Such a culture may well consist of a variety of differentiated cell types. In the case of connective tissue, fibroblasts will start growing from these fragments within about 7 days. Fibroblasts have particularly good growth characteristics and often outgrow other cells which may be present initially. Careful control of the composition of the growth medium may allow the selective growth of certain cell types such as epithelial cells.

Blood contains a range of different cell types in suspension which can be fractionated by density gradient centrifugation. Formulations such as 'Ficoll' and 'Percoll' (from Pharmacia) are available for establishing gradients which are optimal for the separation of these blood cells in sterile medium (see *Protocol 10*, Chapter 6). Such separations have been used extensively for the isolation of single cell types from primary sources (18, 19).

3.2 'Normal' cells

Subculture of a primary culture leads to secondary and tertiary cultures. In the case of suspension grown cells such as lymphocytes this involves dilution at high cell density with fresh growth medium. However, most normal animal cells derived from solid tissue are anchorage-dependent. The designation of cells as 'normal' is based on a set of characteristics outlined in *Table 2*. Such cells grow as a monolayer which eventually can completely cover the substratum on which the cells are growing. At this point (i.e. confluence) subculture involves the removal of cells by trypsinization and re-inoculation into fresh medium. The cells may grow at a constant rate over successive subcultures.

Hayflick and Moorhead (13) studied the growth potential of cells derived from normal human embryonic tissue (designated WI-38). They showed that these cells have a finite growth capacity and can be continuously subcultured for about 50 generations. After this time, which takes several months, the cells enter a senescent phase in which there is an observed decline in cell number.

A finite growth capacity of this type is a characteristic of all cells derived from tissue *in vivo*. The extent of growth is related to the origin of the cells—those derived from embryonic tissue having a greater growth capacity than

Table 2. Typical characteristics of 'normal' animal cells

A diploid chromosomal complement which indicates that no gross genetic changes have occurred
Anchorage-dependence and density inhibition which is an indication of growth control and is shown by cessation of growth as a confluent monolayer on the surface of a growth flask
A finite lifespan which is a reflection of the intrinsic growth regulation of the cells
Non-malignancy of the cells which can be shown by the inability to form tumours following injection of the cells into immunocompromised mice

those derived from adult tissue. The natural ageing of the cell population can be arrested by storage. Thus cells from human embryonic tissue stored in liquid nitrogen after 20 generations of growth retain their growth capacity for a further 30 generations when re-inoculated into culture.

3.3 Transformed cells

Some cells can acquire a capacity for infinite growth and such a population cells is a 'continuous cell line' (*Figure 1*). This requires cells to go througl process called transformation which causes cells to lose their sensitivity many stimuli normally associated with growth control. Loss of anchora dependence and cell density inhibition are typical of such changes. Transf mation is sometimes associated with an altered chromosome pattern. Dipl cells often become aneuploid which indicates chromosomal fragmentation

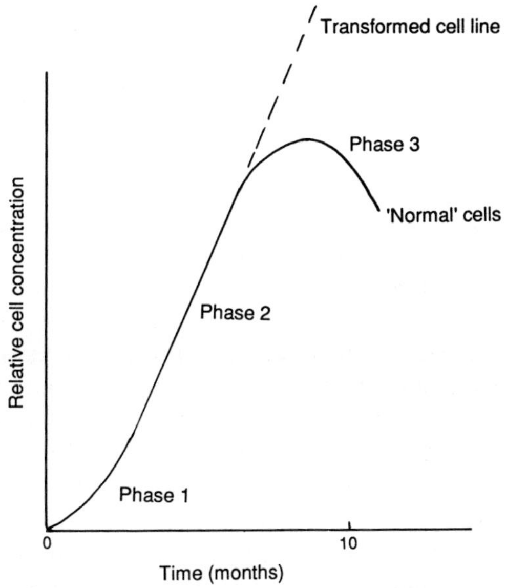

Figure 1. The growth potential of 'normal' and transformed cells in culture.

Cells can be transformed *in vivo* by carcinogenesis which is an analogo but not necessarily identical process. Not all transformed cells are malignan a characteristic defined by the ability to form tumours in immunosuppresse experimental animals. However, many tumour-derived cell lines have bee used extensively in culture. Examples include HeLa cells which are derive from a cervical cancer and Namalwa cells which derived from a huma lymphoma. These cell lines are quite robust. They show good growth chara teristics which include a low doubling time and a low requirement for growt factors.

Some cells exhibit spontaneous transformation in culture. A fraction of a cell population derived from a primary culture may continue growth at the point where most of the cells stop growing. Such spontaneous transformation has been particularly observed with rodent cells and may be explained by the tendency of such cells to carry endogenous viruses. The frequency of the transformation event can be increased by the treatment of cells with a selected virus or mitogen. For example, mouse L cells were originally transformed by treatment with methylcholanthrene.

Although transformed cells are much easier to grow in culture, there has been some reluctance to accept their use for large-scale production of biologicals. The origins of these cells has caused concern over the possibility of contamination of their products with tumourigenic agents. This concern has led to the restriction in the use of transformed cells in large-scale processes particularly for human consumption. However, stringent monitoring and control of the recognized dangers and improved analysis of the final product have more recently led to a relaxation of the restrictions on the use of these cells.

4. Cell lines

Well characterized animal cell lines are often those of choice in the development of production processes. The reasons for this are clear. In order to control productivity it is important to know the growth characteristics of the cells used so that reproducibility can be maintained between batches. Furthermore, if the genetic characteristics of the cell line are known then this allows any unwanted genetic drift to be monitored. Because of this certain well characterized cell lines have emerged as 'favourites' for the animal cell technologist. The characteristics of some of these cell lines are included in the following list.

- *BHK-21*. The Baby hamster kidney cells were derived from five unsexed 1-day-old hamsters in 1961 (20). The widely used clone 13 of these cells was initiated by a single cell isolation in 1963. The original strain was fibroblast-like and anchorage-dependent but late passage number cells can be grown in suspension. The cells have been used extensively for the propagation of viruses including polyoma, foot-and-mouth disease, and rabies, which are used as veterinary vaccines.
- *CHO-Kl*. Chinese hamster ovary cells were derived from an adult Chinese hamster in 1957 (21). The widely used Kl subclone has a requirement for proline because of a genetic deficiency in the conversion of glutamic acid to glutamic γ-semialdehyde. The cells are epithelial-like and can be grown in suspension. These cells have been used extensively for the expression of proteins from recombinant DNA.
- *HeLa*. These cells were isolated from a carcinoma of the cervix of a 31-year-old Negro female (Henrietta Lach) in 1952. This has been one of

the most widely studied human cell lines (22). The cells are epithelial-like, aneuploid, and can be grown in suspension. A number of human cell lines derived in the 1960s were found to be contaminated with HeLa cells. Such cells are available from culture collections and are recognized by the acquisition of genetic markers from HeLa cells.

- *McCoy*. There is little information available as to the origin of these cells (23). The cells now available from culture collections are characterized as anchorage-dependent mouse fibroblasts possessing marker chromosomes similar to mouse L cells.
- *MDCK*. Madin Darby canine kidney cells were derived from a normal adult female cocker spaniel in 1958 (23). Although the original cells appeared to be fibroblast-like, subsequent cloning and subculture resulted in an epithelial-like cell line. The cells are anchorage-dependent and can be grown very successfully on microcarriers. The cells support the growth of numerous viruses and have been used for the production of veterinary vaccines.
- *Mouse L*. This was one of the first cell lines to be established in continuous culture and was originally derived from the connective tissue of a 100 day-old male mouse in 1943. The well used clone 929 was derived in 1948 as a single-cell fibroblast and can be grown in suspension culture (24). The cell line was used extensively throughout the 1950s to develop the techniques of cell culture and for the demonstration of carcinogenesis *in vitro*.
- *MRC-5*. Medical Research Council 5 is a human diploid cell line derived from normal embryonic lung tissue in 1966 and with properties similar to those of WI-38 cells (25). The cells have a growth capacity of 42–6 generations but proliferate more rapidly and are less sensitive to adverse environmental factors than WI-38 cells. The cells are susceptible to a wide range of viruses and have been used for human vaccine production.
- *Namalwa*. These are derived from human lymphoblastoid cells from a patient of the same name suffering from Burkitt's lymphoma. The cells have been used for the large-scale production of α-interferon (26).
- *3T3*. These cells were fibroblasts derived from Swiss mouse embryos in 1963. The name is derived from the procedure of 3 days growth followed by transfer to three subcultures (27). The cell line is anchorage-dependent and exhibits a high degree of contact inhibition. Consequently, the cells have offered a good system for the study of cell transformation by oncogenic viruses.
- *Vero*. The cell line was initiated from the kidney of a normal adult African green monkey in 1962. The cells are anchorage-dependent fibroblasts and grow continuously in culture (28). The cells support the propagation of numerous viruses including poliovirus which has been approved for human use.

- *WI-38*. Wistar Institute 38 is a human diploid cell line derived from normal embryonic lung tissue of a Caucasian female in 1961. These cells have a doubling time of 24 h and a finite lifespan of 50 generations. The cells are fibroblasts and produce collagen. The cells were used to characterize 'normal' cells (13). They have been used to produce human vaccines.

5. Culture media

The media originally used for the growth of animal cells were based entirely on biological fluids such as plasma and embryonic extracts. The use of chemically undefined media, however, suffers from the disadvantages of batch variation and vulnerability to contamination. Chemically defined culture media were originally based upon the analysis of plasma and this led to complex formulations such as Medium 199 which contains over 60 synthetic ingredients.

An alternative approach in media design is to reduce the number of components to the minimum shown to be essential for cell growth. This strategy led to Eagle's basal medium (BME) which was designed for the optimal growth of mouse L cells and HeLa cells (29). It should be noted that there are several versions of BME. The formulation outlined in *Table 3* is widely used and is available commercially. This was later modified and improved (16) as *Eagle's minimum essential medium* (EMEM) which has also found wide application for the growth of a variety of cell lines.

Although there are numerous media formulations now available for cell growth, the list given below and in *Table 3* indicate those which have found greatest application for the cell technologist. The choice of a particular medium for a cell type has to be empirical but those listed are those which would normally be a starting point for attempting the growth of any new cell line.

- *Glasgow's modification of Eagle's medium* (GMEM) is a modification of BME and contains 2 × the concentrations of the amino acids and vitamins with extra glucose and bicarbonate (20). This media was originally developed for the growth of BHK21/C13 cells and is usually supplemented with 10% tryptose phosphate broth as well as serum (30).
- *Dulbecco's modification of Eagle's medium* (DMEM) has a high nutrient concentration and includes 4 × the BME concentration of amino acids and vitamins as well as additional non-essential amino acids and trace elements (31). It was first reported for the culture of embryonic mouse cells but has since found a wide application for the culture of various cells including primary mouse and chicken cells. The original formulation contained 5.6 mM glucose but 25 mM is now normally used.
- *Ham's F12* has a complex composition including various trace elements and was originally designed for cloning diploid hamster ovary cells (32). It was

Table 3. The composition of commonly used media formulations

	BME	EMEM	GMEM	DMEM	F12	RPMI	eRDF	IMDM
Inorganic salts (mM)								
NaCl	117	116	110	110	131	100	110	77
KCl	5.0	5.4	5.4	5.4	3.0	5.4	5.0	4.4
KNO_3								0.75×10^{-3}
$CaCl_2$	1.0	1.8	1.8	1.8	0.3			1.5
$Ca(NO_3)_2$						0.42		
$MgCl_2$	0.5	1.0						
$MgSO_4$			0.8	0.8	0.6	0.41	0.55	0.81
NaH_2PO_4	1.01	1.1		0.9			18.41	1.0
Na_2HPO_4			0.9		1.0	5.64		
$NaHCO_3$	20.0	23.8	32.7	44.0	14.0	23.8	12.5	36.0
Trace elements (μM)								
$CuSO_4$					0.01		3×10^{-3}	
$FeSO_4$					3.0		0.8	
$Fe(NO_3)_3$			0.25	0.25				
$ZnSO_4$					3.0		0.8	
H_2SeO_3								0.1
L-Amino acids and derivatives (mM)								
Alanine					0.1		0.075	0.28
Arginine	0.12	0.6	0.24	0.4	1.0	1.15	2.76	0.4
Asparagine					0.1	0.33	0.63	0.19
Aspartate					0.1	0.15	0.3	0.23
Cysteine					0.2		0.6	
Cystine	0.05	0.1	0.1	0.2		0.21		0.29
Glutamate					0.1	0.14	0.27	0.51
Glutamine	2.0	2.0	4.0	4.0	1.0	2.0	6.84	4.0
Glutathione						3.2×10^{-3}	1.6×10^{-3}	
Glycine				0.4	0.1	0.13	0.57	0.4

Table 3. Contd.

	BME	EMEM	GMEM	DMEM	F12	RPMI	eRDF	IMDM
L-Amino acids and derivatives (mM)								
Histidine	0.05	0.2	0.1	0.2	0.1	0.1	0.36	0.2
Hydroxyproline						0.15	0.24	
Isoleucine	0.2	0.4	0.4	0.8	0.3	0.38	1.2	0.8
Leucine	0.2	0.4	0.4	0.8	0.1	0.38	1.26	0.8
Lysine	0.2	0.4	0.4	0.8	0.2	0.22	1.08	0.8
Methionine	0.05	0.1	0.1	0.2	0.03	0.1	0.33	0.2
Phenylalanine	0.1	0.2	0.2	0.4	0.03	0.09	0.45	0.4
Proline					0.03	0.17	0.48	0.35
Putrescine					1×10^{-3}		0.25×10^{-3}	
Serine				0.4	0.1	0.29	0.81	0.4
Threonine	0.2	0.4	0.4	0.8	0.1	0.17	0.93	0.8
Tryptophan	0.02	0.05	0.04	0.08	0.01	0.02	0.09	0.078
Tyrosine	0.1	0.2	0.2	0.4	0.03	0.11	0.48	0.46
Valine	0.2	0.4	0.4	0.8	0.1	0.17	0.93	0.8
Carbohydrates and derivatives (mM)								
Glucose	5.0	5.6	25	25	10	11.1	19.0	25
Pyruvate				1	1		1	1

Nucleic acid derivatives (μM)								
Hypoxanthine					30		7.5	
Thymidine					3		23.6	
Lipids and derivatives (μM)								
Choline	8.3	8.3	14.3	28.6	100	21.4	88	29
Inositol		11.0	22.2	38.9	100	194.4	260	40
Linoleate					0.3		0.075	
Vitamins/co-factors (μM)								
Aminobenzoic acid						7.3	3.7	
Biotin	4.1				0.03	0.82	4.1	0.053
Folic acid	2.3	2.3	4.5	9.1	3.0	2.27	4.1	9.1
Lipoic acid					1		0.25	
Nicotinamide	8.2	8.2	16.4	32.8	0.3	8.2	12	32.8
Pantothenate	4.6	4.6	9.0	17.0	1.0	1.0	2.6	17.0
Pyridoxal	6.0	6.0	9.8	19.6			4.9	19.6
Pyridoxine					0.3	4.9	2.5	
Riboflavin	0.27	0.27	0.5	1.1	0.1	0.5	0.56	1.1
Thiamine	3.0	3.0	5.9	11.9	1.0	3.0	4.7	11.9
Vitamin B_{12}					1.0	3.7×10^{-3}	0.25	9.6×10^{-3}
Buffers/indicators								
Hepes (mM)							5	25
Phenol red (μM)	14	14	14	12	10	14	14	40
CO_2 in gas phase (%)	5	5	5	10	5	5	5	10

originally designed as a serum-free formulation but now is commonly used with a serum supplement to support the growth of a variety of normal and transformed cells. F12 has been combined with DMEM (1:1) as the basis for development of serum-free formulations (33). The idea was to combine the richness of F12 with the high nutrient concentrations of DMEM.

- *RPMI 1640* was developed at Roswell Park Memorial Institute for the culture of normal human leucocytes (34). This media has been particularly well used for the growth of hybridomas.
- *eRDF* is based on a mixture of RPMI 1640:DMEM:F12 (2:1:1). This has been found to be effective with a serum-free formulation for the support of hybridomas (35). The enriched form of RDF contains 3 × the original concentration of amino acids and 2 × the concentration of glucose. Such a formulation with the addition of ITES (insulin, transferrin, ethanolamine, and selenite) can be more effective for myeloma or hybridoma cell growth than a serum-based medium.
- *Iscove's modified Dulbecco's medium* (IMDM) is a modification of DMEM containing additional amino acids and vitamins, selenium, sodium pyruvate, Hepes, and potassium nitrate instead of ferric nitrate. This medium has proved useful as a basis of serum-free formulations (36) and has been widely used for the growth of lymphocytes and hybridomas (see Chapter 6).

Standard media formulations such as those outlined above can be purchased from commercial suppliers (e.g. Sigma, Flow) as 1 × or 10 × concentrates, the latter requiring dilution with sterile double-distilled water. Alternatively, powdered media may be preferred especially if large quantities are needed. In this case, the powder should be dissolved in distilled water and sterilized by filtration through a 0.22 μm filter under positive pressure with nitrogen. Sterilized supplements of sodium bicarbonate, antibiotics, and glutamine are often added after sterilization of the bulk of the medium. This is essential if the medium is autoclaved, as these supplements are unstable.

The components of these media formulations include a complex mixture of carbohydrate, amino acids, salts, vitamins, hormones, and growth factors. The carbohydrate included is normally glucose which was originally thought to be the major energy source. However, evidence now shows that the amino acid carbon skeletons are also important as energy sources—particularly glutamine which is included at high concentrations relative to the other amino acids (2–4 mM). In most cultures, glutamine and glucose are utilized particularly rapidly and can cause cell growth limitations even before their complete exhaustion.

The media salt concentration is isotonic to prevent osmotic imbalances. The osmolality of standard growth media is approximately 300 mOsm/kg and is optimal for most cell lines. Care should be taken when adding supplements to the media as this may well alter the osmolality. However, many cell lines have been shown to tolerate variations of approximately 10% of this optimal value.

Bicarbonate is often included to act as a buffer system in conjunction with the carbon dioxide environment (5–10%) in which the cells should be cultured. This allows the cultures to be maintained at the normally optimum pH range of 6.9–7.4. The CO_2 can be provided by a controlled atmosphere in an incubator or, in a sealed small flask, sufficient CO_2 may be generated by the culture. The disadvantage of the bicarbonate–CO_2 buffer system is that the cultures may become alkaline very quickly when removed from the CO_2 incubator. To prevent this the organic buffer Hepes (pK 7.0) may be added at a concentration up to approximately 25 mM. This allows the CO_2 level to be reduced to 2%.

The vitamin and hormone components are present at relatively low concentrations and are utilized as co-factors, the requirements for which show considerable variations between cell lines. Thus, the content of these co-factors varies considerably among different media formulations.

Antibiotics are often included in media for short-term cultures. The use of antibiotics for routine subculture or in stock cultures should be avoided as this may mask a low-level contamination which may cause problems at a later date. Furthermore, extensive use of antibiotics may cause the selective retention of antibiotic-resistant contaminants in the laboratory. When antibiotics are used, the following cocktail may be recommended:

* penicillin (100 IU/ml) to inhibit the growth of Gram-positive bacteria
* streptomycin (50 μg/ml) to inhibit the growth of Gram-negative bacteria
* amphotericin B (25 μg/ml) as an anti-fungal agent.

For most cultures, supplementation with blood serum at 10% v/v is required to maintain cell growth. The serum can be obtained from various animal sources—bovine or equine being the most common. One of the most effective supplements for cell growth is fetal calf serum because of its high content of embryonic growth factors. However, the inclusion of serum in growth medium has many disadvantages. It is a chemically undefined mixture and variation between batches can result in inconsistent results in cell growth. Serum supplementation can be expensive and can account for 70 to 80% of the overall medium cost if fetal calf serum is used. Furthermore, extracellularly released products are mixed into a soup of proteins from which it is difficult to extract.

For these reasons there has been considerable effort made in establishing low-serum or serum-free formulations for growing cells (see Chapters 2 and 3 and reference 7). Such serum substitutes can include defined hormone cocktails such as the well used HITES or ITES media which contain hydrocortisone, insulin, transferrin, ethanolamine, and selenite. Alternatively, serum-free formulations may include growth factor extracts from endocrine glands such as epidermal or fibroblast growth factors. Every cell line is specific in its growth requirements and each serum-free formulation can only be applied to a limited number of cell types. Serum-free formulations have worked well for fast growth tumourigenic or transformed cell lines, whereas their development for certain fastidious cell lines can be more difficult.

6. Cell growth

Following cell inoculation, growth follows the pattern typically found for micro-organisms (*Figure 2*). A lag phase may occur, the length of which is dependent upon the state of the cells at inoculation. Cells taken from an exponentially growing culture exhibit a short lag phase on subculture, whereas cells which have been stored or taken from a stationary culture may exhibit an extended lag phase.

The cell density at inoculation also influences the extent of the lag phase. A minimum cell inoculum of 10^4–10^5 cells/ml is normally required for early initial growth. Cells are capable of releasing growth factors into the medium and such factors may be required to reach optimal concentrations before growth occurs. At low cell densities of inoculation, such as required for cloning, a feeder layer of cells may be used. Such a feeder layer consists of irradiated cells which are incapable of growth but are metabolically active and capable of releasing growth factors into the medium. Transformed cells have a relatively low requirement for growth factors and may be inoculated at lower cell concentrations.

6.1 Expression of growth data

Exponential growth of cells can be represented by

$$N = N_0 \cdot 2^x$$
$$\text{or } \log_{10} N = \log_{10} N_0 + x \cdot \log_{10} 2$$

where N is the final cell number, N_0 the initial cell number, and x the number of generations of exponential growth.

Thus, a cell inoculum of 10^5 cells/ml which has increased to a cell density of 10^6 cells/ml passes through 1/log 2 or 3.32 generations. The doubling time during exponential growth can be calculated from the equation

$$\text{Doubling time} = \text{total time elapsed/number of generations.}$$

If the above example is taken as a typical culture, then the final cell density of 10^6 cells/ml would normally occur after about 3 days. This would mean that the doubling time is 0.904 days or 22 h.

The specific growth rate (μ) is a term which relates the cell number (or biomass) increase at a certain cell concentration. Thus

$$\mu = \cdot \frac{dN}{dt} \cdot \frac{1}{N}$$

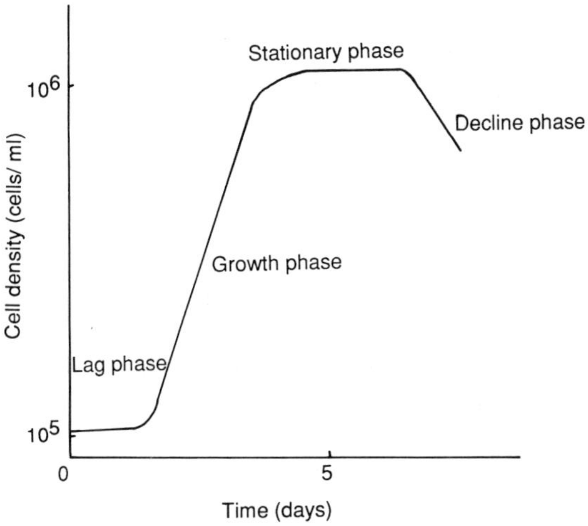

Figure 2. The pattern of animal cell growth in culture.

Integration gives

$$\ln N = \ln N_0 + \mu \cdot t.$$

The value μ (h^{-1}) can be calculated from the slope of the plot of ln N against time, t. For the example above the specific growth rate is 0.032/h.

The exponential growth phase normally exhibits a doubling time of between 15 to 25 h for animal cells. Growth continues in batch culture to $1\text{–}2 \times 10^6$ cells/ml which is the typical cell density sustainable by presently available media formulations. At this point the depletion of nutrients or the accumulation of inhibitory metabolites may be responsible for the cessation of cell growth.

The specific rate of consumption of a substrate or production of a product can be calculated from the equation

$$\frac{dC}{dt} = K.N_0\, e^{\mu.t}$$

where C is the concentration of substrate or product at time t, K the specific rate of consumption or production, and μ the specific growth rate. This equation can be simplified over a sustained period of exponential growth to

$$K = \frac{\Delta C}{T}\left(\frac{\ln N - \ln N_0}{N - N_0}\right)$$

where ΔC is the total concentration change over the elapsed time period, T.

The specific rate of consumption or production (K) is usually expressed as pmol/cell per h or μmol/10^6 cells per h. The term is widely used in the analysis of media changes during culture. Another similar term is the growth yield which is the measure of the number of cells generated following the consumption of a millimole of substrate.

The passage number of a cell population denotes the number of subcultures performed after the original isolation of the cells from a primary source. The passage number and the generation number should be carefully recorded for a cell line which is continuously subcultured. They may be used to identify points at which changes occur to the cell population during subculture. The relationship between the passage number and the generation number depends upon the split ratio which is the number of new cultures established at each stage of subculture. The simplest case occurs where a confluent culture is subcultured into two new cultures, i.e. a split ratio of 2. In this case the generation number and passage number are the same. Otherwise,

$$\text{Generation number} = \text{passage number} \times \text{split ratio}/2.$$

7. Maintenance of cells in culture

On the laboratory scale, a variety of sterilizable containers are available for cell growth. These include multi-well plates, Petri dishes, T-flasks, Roux bottles, and spinner flasks. These are suitable for growth of both suspension and anchorage-dependent cells.

Cells are routinely propagated by subculture which, in the case of cells grown in suspension, can be performed by dilution. Thus, cells from the stationary phase of one culture can be re-suspended in fresh medium at 2×10^4 to 5×10^4/ml for fast growing cells (12–18 h doubling time) or approximately 10^5/ml for slow growing cells (24–48 h doubling times).

7.1 Culture of anchorage-dependent cells

Anchorage-dependent animal cells require a substratum for cell growth and form a monolayer of cells covering the available surface. For such cells the interaction between the cell membrane and the solid surface is critical. Adhesion between the solid surface and the negatively charged cell membrane is maintained by the involvement of divalent cations (normally Ca^{2+}) and basic proteins which form a layer between the substratum and the cells. Fibronectin is found on the surface of many cells and has been implicated as a specific cell adhesion molecule, particularly if collagen is present on the substratum. However, in most cases of cell–substratum interaction adhesion is provided by a range of non-specific proteins. The adhesive interaction involves a combination of electrostatic attraction and van der Waal's forces. Tissue-grade plastic normally consists of sulphonated polystyrene which has a negatively charged surface density which is optimized at 2–6 groups/nm^2. The negative charge on the glass surface of containers can be provided by alkali treatment.

The subculture of anchorage-dependent cells requires their detachment from the substratum prior to re-inoculation in a fresh culture. Trypsin is the most commonly used proteolytic enzyme for cell detachment although alternatives include Pronase (Sigma) and Dispase (Boehringer).

Protocol 1 has been found suitable for the subculture of 'robust' cells such as MDCK or BHK cells. If the treatment is found to be too harsh for particular cells resulting in poor viability, then a gentler trypsinization treatment may be suggested. One such method involves exposing the cells to the reagent for only 15–30 sec at 4°C before decanting. Then allow the cells to detach in the residue of trypsin by incubating for 5–15 min.

The surface area of the substratum may limit the growth of anchorage-dependent cells. Such cells do not normally form multilayers and so growth ceases if the available substratum is covered by a monolayer of cells. Cell densities obtainable in a monolayer vary with the type of cells and type of substratum but would normally be expected to be within the range of 5×10^5 cells/cm^2. Thus a 200 cm^2 Roux bottle could produce approximately 10^8 cells.

Protocol 1. The subculture of anchorage-dependent cells

The following protocol is described for the subculture of 'robust' anchorage-dependent cells into 150 cm^2 T-flasks.

Materials

- Phosphate buffered saline (PBS) (0.1 M NaCl, 8.5 mM KCl, 0.13M Na_2HPO_4, 1.7mM KH_2PO_4)
- Dissociating reagent (1% trypsin + 2% EDTA in PBS)

Method

1. Measure 40–50 ml of growth medium into a series of new T-flasks and label with the cell line name, passage number, and date. Incubate the flasks at 37°C while the cells are being trypsinized.
2. Remove the medium from a pre-confluent flask of cells. If cells are harvested at confluence, there may be some interruption of cell growth which may result in a lag phase on subculture.
3. Rinse the flask with 20 ml of PBS. The purpose of this is to remove any remaining medium-derived protein which would act as a substrate for trypsin.
4. Decant off the PBS and remove the last drops with a pipette.
5. Add 5 ml of the dissociating reagent which is a mixture of EDTA and trypsin. Incubate the flask at 37°C for 10 min. The flask may be shaken vigorously at periodic intervals to accelerate the removal of the cells from

Protocol 1. *continued*

the surface of the flask. The progress of cell detachment can be followed by visual observation.

6. When the cells are detached, add 8 ml of serum-supplemented growth medium to the flask. Trypsin inhibitors in the medium will prevent excessive damage to the cells by trypsin.
7. Centrifuge at low speed and re-suspend the cells in 20 ml of medium.
8. Aliquot the cell suspension into the pre-prepared flasks at the desired split ratio. A 1:5 split should allow the subcultured cells to reach confluence in about 3 days.

7.2 Microcarrier cultures

Microcarriers are charged microscopic particles (beads of 200 μM diameter) which can act as a substratum for cell attachment and growth (37, 38). Cells are inoculated into the growth medium containing the microcarriers which can be maintained in suspension by gentle stirring. The culture conditions are relatively homogeneous and are sometimes referred to as 'pseudo-suspension'. The advantage of such cultures is that a large surface area is offered for cell growth. Thus a 1 litre suspension of microcarriers can offer the same growth surface as 20 roller bottles.

A large range of microcarriers are commercially available (*Table 4*). The most widely used standard is Cytodex 1 (Pharmacia) which consists of dextran

Table 4. Microcarriers which are commercially available

Type	Trade name	Source	Composition
Dextran	Cytodex 1	Pharmacia (Sweden)	DEAE–dextran
	Cytodex 2	Pharmacia (Sweden)	Quaternary amine-coated dextran
	Dormacell	Pfeifer & Langen (Germany)	DEAE dimers–dextran
Plastic	Biocarriers	Biorad (USA)	Polyacrylamide/DMAP
	Cytospheres	Lux (USA)	Polystyrene—charged
	Micarcel G	Réactifs IBF (France)	Polyacrylamide/collagen/ glucoglycan
	Bioplas	Solohill Eng. (USA)	Polystyrene—crosslinked
	Mica	Muller-Lierheim (Germany)	Epoxy resin
Gelatin	Geli-heads	KC Biologicals/ Hazelton Labs (USA)	Gelatin
	Ventregel	Ventrex (USA)	Gelatin
	Cytodex 3	Pharmacia (Sweden)	Gelatin-coated dextran
Glass	Bioglas	Solohill Eng. (USA)	Glass-coated plastic
	Ventreglas	Ventrex (USA)	Glass-coated plastic

beads charged with diethylaminoethyl (DEAE) groups. The density of surface charge is critical for cell attachment and these beads have been optimized for use of a large number of cell types.

Protocol 2 has been found suitable for the culture of a number of robust anchorage-dependent cell lines. The initial attachment of cells to microcarriers is a critical stage because each bead supports the growth of an independent cell colony. After attachment bead-to-bead transfer of cells is rarely found. Therefore, the initial cell inoculum/bead ratio is important in order to establish an even distribution of the cell inoculum on the available beads. A critical ratio has been determined as greater than 7 for MDCK cells. Thus, a cell-to-microcarrier inoculation ratio above the critical value ensures:

(a) a low proportion of unoccupied microcarriers in culture (<5 per cent)

(b) maximum use of the available surface area by the cells

The inoculation concentration suggested in *Protocol 2* calculates to a cell to microcarrier inoculation ratio of 16, i.e. well above the critical value. After a few days growth in such a culture, the surface of the microcarriers would be completely covered by cells. If the microcarrier concentration at inoculation were increased to 5 mg/ml, this would still be above the critical ratio but in batch culture this would lead to incomplete cover of the microcarrier surface. However, the extra availability of surface area for cell growth could lead to increased cell yields in batch-fed or perfusion cultures (see Chapter 7). Microcarrier cultures have been scaled-up for commercial processes using simple stirred tank reactors as described in Chapter 7.

An increased surface area for growth may also be provided by the use of porous particles which have hollow channels within each sphere. These may be of similar size to the surface-based microcarriers, e.g. Cultispher G (Percell Biolytica, Sweden) or may be larger microspheres such as the VX-100 collagen particles (diameter 600 μm) produced by Verax (USA) or porous glass spheres (Siran from Schott, Germany). The larger microspheres are used in fluidized-bed reactors which support the particles by an upward flow of medium.

Protocol 2. The establishment of microcarrier cultures

The following protocol has been found suitable for the establishment of 100 ml microcarrier cultures in spinner flasks.

1. Add 1 g of dry Cytodex microcarriers (Pharmacia) to 100 ml of PBS and leave at room temperature for 3 h with gentle agitation. This allows the hydration of the dextran microcarriers.
2. Discard the PBS and replace with fresh PBS (100 ml).
3. Autoclave the suspension of microcarriers at 115°C for 15 min at 15 psi.

Protocol 2. *continued*

4. Decant the supernatant from the sterilized microcarriers and wash in warm culture medium.
5. Allow the microacarriers to settle and remove the medium.
6. Add 50 ml of fresh medium to the microcarriers (concentration 20 mg/ml).
7. Transfer 15 ml of this suspension to each 250 ml spinner flask.
8. Add a further 75 ml of medium to each flask.
9. Each spinner flask is now ready for the addition of cells which can be added in a 10 ml suspension to make a total culture volume of 100 ml. The inoculation of 3×10^7 cells will allow cell attachment to most of the microcarriers.
10. Place the spinner flasks on a microcarrier stirrer base (Bellco or Techne) and stir at 40–50 r.p.m. A confluent cover of the microcarriers will occur in 3–4 days.

8. Cell counting

The best method of determining the growth of cells in culture is by direct counting of cell samples taken at time intervals—typically every 24 h. The two methods used routinely are microscopic counting or electronic particle counting.

8.1 Haemocytometer

This involves counting the number of cells or nuclei in a pre-determined volume by direct microscopic examination (39). The haemocytometer slide has a pattern of grids which indicates a standard volume under the cover-slip (*Figure 3*). In the widely used Neubauer haemocytometer, there are nine large squares each of which has a volume of 0.1 mm^3 (= 10^{-4} ml). It is common practice to count the cells in five such squares (n). Thus,

$$\text{Number of cells in suspension} = n \times 10^4/5 \text{ (/ml)}.$$

A count of viable cells can be made by prior incubation with a suitable dye (40) such as trypan blue (0.2%). Viable cells are able to exclude the dye whereas non-viable cells are stained blue. Alternatively, stained nuclei may be counted (*Protocol 3*; reference 41). The latter method is particularly suitable for counting anchorage-dependent cells attached to a surface.

Protocol 3. Counting stained nuclei

This technique can be used for counting cells in suspension or for those bound to a surface such as microcarriers. The principle of the method is that a

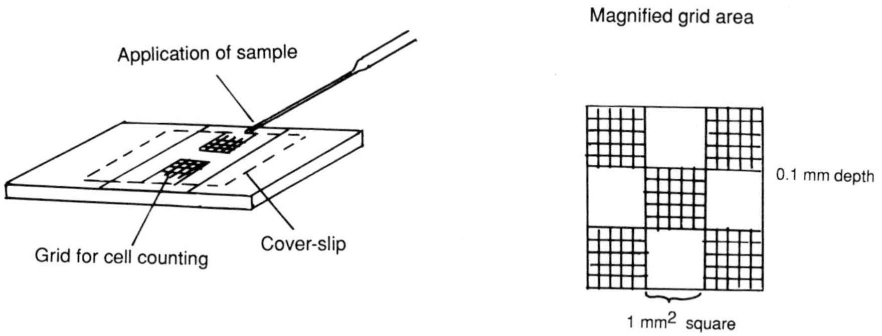

Figure 3. The haemocytometer for cell counting.

hyperosmotic solution causes the cells to lyse and the released nuclei are stained with crystal violet.

1. Pipette a 1 ml aliquot of a cell suspension into a tube.
2. Add a 1 ml solution of crystal violet (0.1%) + citric acid (0.1 M).
3. Incubate for at least 1 h at 37°C.
4. Introduce a sample by Pasteur pipette into a haemocytometer grid and count the nuclei which are stained purple.

8.2 Coulter counter

The principle of an electronic cell counter (or Coulter counter) is that a predetermined volume of cells in suspension is allowed to pass through a suitable sized orifice and then between two electrodes (42) (*Figure 4*). The passage of each cell interrupts an electrical current and the resulting voltage change is recorded as a signal on an electronic counter.

The major advantage of this method is the speed of analysis and is therefore suitable for counting a large number of samples. However, the method is based upon the number of particles contained in suspension and, consequently, the proportion of viable cells in the sample can not be determined. Also, it must be ensured that cell aggregates are not present in the sample; otherwise, the actual cell count will be underestimated.

9. Storage of cells

Cells can be stored for long periods at low temperatures. This permits cell stocks to be maintained for culture inocula without having to resort to primary tissue. The maintenance of a master cell stock also guards against loss of a cell line by contamination or genetic change which may occur by continuous subculture. It is often useful to store such cells at various passage numbers.

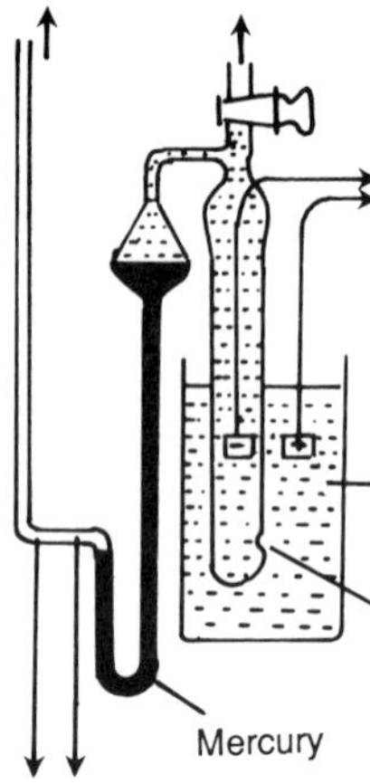

Figure 4. An electronic cell counter (Coulter).

This allows the monitoring of any genetic change that may have occurred during the culture period.

Cells can be stored for long periods without deterioration in liquid nitrogen (−196°C). However, in order to prevent damage to the cell membrane by ice crystal formation, the cells are suspended in a cryoprotective agent. This normally consists of growth medium supplemented with 5% dimethylsulphoxide (DMSO) or 10% glycerol. DMSO is often preferred because of the greater rate of penetration into the cells. However, the exposure of cells to DMSO above freezing temperatures should be brief. For sensitive cell lines fetal calf serum supplemented with DMSO (5%) can be used. Slow freezing and rapid thawing is recommended to maintain high cell viability (*Protocols 4* and *5*).

Protocol 4. Freezing cells for storage

1. Suspend the cells in the desired cryopreservative at a concentration of 5×10^6–2×10^7 cells/ml.
2. Dispense the cell suspension into plastic ampoules (1.5 ml) with screw caps. These are specifically designed for cell storage.
3. The optimal freezing rate is 1°C/min. To achieve this, simply place the vials of cells into a polystyrene box of 1.5 cm wall thickness at −70°C for 2 h before transfer to liquid nitrogen. This freezing rate may be more carefully controlled by use of a programmable cooler.

Protocol 5. Cell recover after storage

1. Thaw each vial rapidly by placing in a water bath at 37°C.
2. Swab the outside the of each vial with 70% ethanol before opening under sterile conditions in a laminar flow cabinet.
3. Transfer the cell suspension into some pre-warmed growth medium (10 ml for 10 × dilution of cells). For sensitive cell lines, the medium should be added slowly to the cells at a rate of about 5 ml/min.
4. Re-suspend the cells in fresh medium as soon as possible to remove the cryopreservative.

9.1 Cell banks

Cell lines may be obtained from cell culture collections (banks) which have a large selection of well characterized cell lines available for purchase. The full history of each cell line is available as well as the criteria used to determine its authenticity. Other services offered by these establishments include:

(a) safe depository of private cell collections. This is useful for maintaining a master stock of valuable cell lines from an individual laboratory.

(b) testing for contamination in cell lines. This includes bacterial, fungal, viral, and mycoplasma testing.

(c) characterization of cell lines. This includes isoenzyme analysis and chromosome analysis. Also, more recently DNA fingerprinting has been included as a technique for characterization.

The two most important collections of cell lines for animal cell technologists are:

The American Type Culture Collection (ATCC),
12301 Parklawn Drive,
Rockville, Maryland,
USA.

The European Collection of Animal Cell Cultures (ECACC),
Public Health Laboratory Service (PHLS),
Centre for Applied Microbiology Research (CAMR),
Porton Down,
Salisbury SP4 0JG, UK.

References

1. Paul, J. (1975). *Cell and Tissue Culture*. Churchill Livingstone, Edinburgh.
2. Sharp, J. A. (1977). *An Introduction to Animal Tissue Culture*. E. Arnold, London.
3. Jakoby, W. B. and Pastan, I. H. (ed.) (1979). *Methods in Enzymology: Cell Culture*, Vol. 58. Academic Press, New York.
4. Adams, R. L. P. (1980). *Laboratory Techniques in Biochemistry and Molecular Biology: Cell Culture for Biochemists*. Elsevier, Amsterdam.
5. Barnes, D. W., Sirbasku, D. A., and Sato, G. H. (ed.) (1984). *Cell Culture Methods for Molecular and Cell Biology*, Vols 1–4. A. R. Liss, New York.
6. Taub, M. (ed.) (1985). *Tissue Culture of Epithelial Cells*. Plenum, New York.
7. Freshney, R. I. (ed.) (1986) *Animal Cell Culture: A Practical Approach*. IRL Press, Oxford.
8. Spier, R. E. and Griffiths, B. (ed.) (1986–90). *Animal Cell Biotechnology*, Vols 1–4. Academic Press, London.
9. Thilly, W. G. (ed.) (1986) *Mammalian Cell Technology*. Butterworths, Boston.
10. Freshney, R. I. (1987). *Culture of Animal Cells: A Manual of Basic Techniques*. A. R. Liss, New York.
11. Butler, M. (1987). *Animal Cell Technology: Principles and Products*. Open University Press, Milton Keynes.
12. Harrison, R.G. (1907). *Proc. Soc. exp. Biol. Med.* **4,** 140.
13. Hayflick, L. and Moorhead, P. S. (1961), *Exp. Cell Res.* **25,** 585.
14. Mascona, A. A. and Mascona, M. H. (1952). *J. Anat.* **86,** 287.
15. Earle, W. R., Bryant, J. C., Schilling, E. L., and Evans, V. S. (1956). *Ann. NY Acad. Sci.* **63,** 666.
16. Eagle, H. (1959). *Science* **130,** 432.
17. Enders, J. F., Weller, T. H., and Robbins, F. C. (1949). *Science* **109,** 85.
18. Pertoft, H. and Laurent, T. C. (1977). In *Methods of Cell Separation* Vol. 1 p. 25 (ed. N. Catsimpoolas). Plenum Press, New York.
19. Sykes, J. A., Whitescarver, J., Briggs, L., and Anson, J. H. (1970). *J. natl Cancer Inst.* **44,** 855.
20. Stoker, M. G. P. and MacPherson, J. A. (1964). *Nature* **203,** 1355.
21. Puck, T. T. (1958). *J. exp. Med.* **108,** 945.
22. Gey, G. O., Coffman, W. D., and Kubicek, M. T. (1952). *Cancer Res.* **12,** 364.
23. American Type Culture Collection (1988). *Catalogue of Cell Lines*. American Type Culture Collection, Rockville, Maryland.
24. Sanford, K. K., Earle, W. R. and Likely, G. D. (1948). *J. natl Cancer Inst.* **9,** 229.
25. Jacobs, J. P. (1970). *Nature* **227,** 168.
26. Finter, N. B. (1973). *Int. J. Cancer* **12,** 396.
27. Todaro, G. and Green, H. (1963). *J. Cell Biol.* **17,** 299.
28. Yasumara, Y. and Kawakita, Y. (1963). *Nippon Rinsho* **21,** 1209.
29. Eagle, H. (1955). *J. Biol. Chem.* **214,** 839.
30. MacPherson, I. A. and Stoker, M. G. P. (1962). *Virology* **16,** 147.
31. Dulbecco, R. and Freeman, G. (1959). *Virology* **8,** 396.
32. Ham, R. (1965). *Proc. natl Acad. Sci.* **53,** 288.
33. Barnes, D. and Sato, G. (1980). *Cell* **22,** 649.

34. Moore, G. E., Gerner, R. E., and Franklin, H. A. (1967). *J. Am. med. Assoc.* **199,** 519.
35. Murakami, H. (1989). In *Advances in Biotechnological Processes*, Vol. 11 (ed. A. Mizrahi), p. 107. A. R. Liss, New York.
36. Iscove, N. and Melchers, F. (1978). *J. exp. Med.* **147,** 923.
37. Butler, M. (1988). In *Animal Cell Biotechnology*, Vol. 3 (ed. R. E. Spier and J. B. Griffiths), p. 284. Academic Press, London.
38. Butler, M. (1987). *Adv. biochem. Eng.* **34,** 57.
39. Patterson, M. K. (1979). In *Methods in Enzymology*, Vol. 58 (ed. W. B. Jakoby and I. H. Pastan), p. 141. Academic Press, New York.
40. Phillips, H. J. (1973). In *Tissue Culture: Methods and Applications* (ed. P. F. Kruse Jr and M. K. Patterson), p. 406. Academic Press, New York.
41. Sanford, K. K., Earle, W. R., Evans, V. J., Waltz, H. K., and Shannon, J. E. (1951). *J. natl Cancer Inst.* **11,** 773.
42. Coulter, W. H. (1956). *Proc. Soc. natl Elect. Conf.* **12,** 1034.

2

Serum and its fractionation

ALEX J. MACLEOD

1. Introduction

Media for maintenance of animal cells *in vitro* are usually aqueous solutions of a number of chemically defined low-molecular-weight components such as amino acids, vitamins, and salts, to which it is normally necessary to add a protein supplement. Animal blood serum is a well established medium supplement being a multifunctional reagent satisfying both nutritional and environmental requirements of the cells. The functions of the proteins in the serum include:

(a) attachment and spreading of adherent cell lines
(b) nutrition (trace elements, labile or water-insoluble nutrients)
(c) stimulation (growth factors, hormones, proteases)
(d) protection:
   (i) biological (antitoxin, antioxidant, antiprotease)
   (ii) mechanical in agitated systems

Serum is derived from the extracellular plasma which constitutes about 80% of the volume of whole blood and contains several hundred different proteins with a total protein content of about 70 g/litre; the major protein components of plasma are listed in *Table 1*. The sera of many species of animals are used as supplements for cell culture media but those most widely used are fetal and new-born bovine and adult horse sera. Each of these or any other serum has its advantages and disadvantages which may be particular to a given application and which must be carefully considered before selecting one for use. Thus, fetal bovine serum is particularly potent in promoting the growth of many types of cells and contains very low levels of immunoglobulins; however, it is also expensive and has been found to have considerable batch-to-batch variation in its composition which is reflected in its cell growth promoting properties. A major consideration in the use of new-born or adult bovine sera collected at abattoirs from intensively reared animals is the probability that they will be contaminated with infectious viruses (especially the bovine diarrhoea virus) and that it is not possible to inactivate these viruses in whole serum.

Table 1. Major protein components of plasma

Protein	**Approximate quantity in normal human plasma (g/litre)**
Albumin	35–55
Immunoglobulin G	8–18
Immunoglobulin A	0.8–4.5
Fibrinogen	3
Transferrin	3
α_1-antiprotease	3
α_2-macroglobulin	2.5
Immunoglobulin M	0.6–2.5
Haptoglobin	1.7–2.3
Fibronectin	0.3

It has been clear for a long time that it is necessary to optimize the formulation of the defined components of media to be used for different types of cells, but only relatively recently has substantial progress been made in identifying the function of the various constituents of serum and optimizing their formulation for *in vitro* cultivation of different types of cells. A requirement for a particular protein in the medium varies with the type of cell to be cultured. Some cell lines such as Namalwa can be grown satisfactorily in medium in which the only protein is albumin. Many other cell lines have been found to require several supplementary proteins, typically including albumin, transferrin, and insulin with the possible addition of polypeptides growth factors, which may be isolated from non-serum sources and which have been shown to stimulate a variety of cell types in culture (see Chapter 3).

The desire to understand the basis of the control of cell growth and to avoid problems introduced by the use of serum has resulted in the development of many 'serum-free' media. These media have the advantage that they can be prepared to standard formulations using highly purified proteins which may be treated to inactivate any viruses present in them. However, cell lines are often very specific in their requirements for growth in serum-free media, and there have been reports that the stability and productivity of some cell lines is reduced in these media. On the other hand serum is very versatile and will support the growth of many unrelated cell lines in broadly similar media with good cell line stability and productivity. Serum-free media are also often more expensive than serum-supplemented media. The fact that cells can thrive in low-protein media demonstrates the high degree of redundancy in the protein content of serum. Thus, selective fractionation of serum or plasma results in medium supplements which may be treated to inactivate contaminating viruses and which support good cell growth.

Human plasma protein fractions are produced in substantial quantities as a result of routine production of therapeutic blood products. At least one of these fractions, Cohn fraction IV, contains high levels of albumin, transferrin, and insulin-like growth factors. This fraction is normally discarded but it has been found that it can be processed to make a low-protein medium supplement capable of supporting growth of a wide-range of cells as a direct replacement for fetal calf serum (1). It is important to appreciate that preparations from plasma contain proteins in the form in which they circulate in the normal healthy individual. In contrast, preparations from serum contain proteins that will have been activated or changed by proteolytic activity during the clotting process. In practice, we have found that plasma protein preparations are as potent as serum proteins in supporting cell growth and protein production, but that a period of adaptation, which may take several weeks, is required for cells to grow in the replacement supplement. This may reflect differences in the activation state of the proteins.

2. Cohn cold ethanol fractionation of serum or plasma

2.1 The principles of Cohn fractionation

Edwin Cohn developed a system of fractionation of human blood plasma in the 1940s to produce a solution of albumin for use as a volume expander in cases of severe loss of circulating fluid. The process that he developed uses selective precipitation of proteins from plasma with cold ethanol and is the basis of an industry which processes in excess of one million kilograms of plasma annually throughout the world.

Cohn identified five variables that could be used to precipitate sequentially the different types of protein from the plasma. These were:

- protein concentration
- ethanol concentration
- pH
- ionic strength
- temperature

To these variables 'time' must be added to allow for the kinetic aspect of the process.

With plasma feedstock it is not generally practicable to adjust protein concentration or ionic strength for bulk processing. Ethanol has a significant positive heat of solution in aqueous mixtures and, consequently, to dissipate the heat generated by the addition of ethanol to the plasma as well as to minimize bacterial growth, the temperature must be kept as low as practicable, certainly below 0°C. This leaves ethanol concentration and pH that can be manipulated to precipitate the proteins and the process can be described as a series of pH/ethanol concentration combinations (2) each of which is associated with a particular fraction that can be characterized by its major components. The most typical version of the process is based on Cohn's method six and generates five fractions (*Table 2*).

Table 2. Cohn fraction precipitation: conditions and composition

Fraction	pH	Ethanol content (% v/v)	Major protein constituents
1	7.2	8	Fibrinogen, Fibronectin
2 + 3	6.7	22	2 Immunoglobulins 3 α and β globulins
4	5.2	20	α and β globulins
5	4.8	30	Albumin

The supernatant to fraction 5 contains less than 2% of the protein in the original plasma. The precipitates generated by this process are not purified proteins, and the early ones in particular contain a broad cross-section of plasma proteins enriched with their major components.

2.2 Production of Cohn fractions for use in cell culture media

Cold ethanol precipitation can be used directly for the production of an improved cell culture medium supplement by precipitation of the large molecular weight proteins including the immunoglobulins as a fraction 1 + 2 + 3 precipitate (*Protocol 1*). Critical features of this process are to keep the temperature as low as possible, adjust the pH, and then add the ethanol slowly with thorough mixing. On a small scale the plasma or serum may be cooled by immersing the beaker in a freezing mixture and ensuring that the temperature does not rise above 0°C but, if at all possible, a refrigerated environment at −5°C should be used and the plasma kept at that temperature throughout the process.

Protocol 1. Production of supernatant to Cohn fraction 1 + 2 + 3

1. Adjust the temperature of the serum or plasma to between 0 and −5°C and keep it at that temperature.
2. Adjust the pH to 6.7 by slow addition of 4.8 M acetate buffer pH4 with constant mixing.
3. Add ethanol slowly to 22% (v/v) with constant mixing.
4. Mix the suspension for several hours, preferably overnight.
5. Collect the precipitate by centrifugation at 4000 *g*.[a]
6. Diafilter the supernatant with five volumes of phosphate buffered saline against a filtration membrane having a 10 000 nominal molecular weight limit pore size.
7. Concentrate the solution to give a total protein content of 100 g/litre.
8. Polish the protein solution by passage through 8, 1.2, and 0.45 μm pore size filtration membranes.

9. Sterilize the protein solution by passage through a 0.2 μm pore size filtration membrane.

[a] Fraction 4 can be precipitated by repeating steps **1** to **5** but taking the pH to 5.2 and the ethanol content to 20% (v/v).

The extensive diafiltration to which this product is subject removes practically all of the low-molecular-weight components. Consequently, several supplements are required to ensure that the product will support cell growth (*Table 3*). It has also been found to be valuable to supplement the product with peptone at 3 g/litre but, as the cell growth enhancing properties vary from batch to batch even from the same supplier, it is important to get samples for evaluation before any purchase is made.

The use of fraction 1 + 2 + 3 supernatant as a cell culture medium supplement was developed at the Centre Regional de Transfusion Sanguine, Lille, France.

Table 3. Additions required to fraction 1 + 2 + 3 supernatant

Biotin	13 ng/litre
Sodium selenite	17.3 ng/litre
Vitamin B_{12}	13 ng/litre

3. Fractionation of serum or plasma by polythylene glycol precipitation of proteins

An alternative precipitant that has been used to prepare adult serum for use as a cell culture medium supplement is polyethylene glycol (PEG) of mean molecular weight 4000 or 6000. This precipitant has the advantage that it is completely innocuous as far as the cells in the culture are concerned and the process can be performed at normal room temperatures of about 20 °C (3).

The precipitant is best prepared as a 50% (w/v) solution in water and in this form it can be safely autoclaved. Addition of the PEG solution to serum to give a final concentration of 10% (w/v) precipitates high-molecular-weight components of the serum including the immunoglobulins. The PEG solution should be added slowly and steadily with constant thorough mixing. The suspension should be mixed for at least 1 h after all the PEG has been added and the precipitate can then be collected easily by low-speed centrifugation and the supernatant sterilized by filtration through 0.2 μm. The freshly prepared product can be used quite satisfactorily as a culture medium supplement; however, on storage a further precipitate is formed which must be re-dissolved by warming the serum carefully to about 35 °C before use. The precipitate will form if the serum is stored at 4 °C but is most marked if it is frozen and thawed when complete resolution may not be possible. The formation of a precipitate can be reduced if the serum is diluted to give a PEG concentration of 7.5% but a corresponding adjustment must be made in the volume of serum used to supplement the medium. Another major problem with this product is that it is not possible to heat it at all to inactivates any infectious micro-organisms present as gross denaturation of the proteins will occur.

4. Subfractionation of bulk Cohn fractions

4.1. Problems associated with Cohn fractions

The human plasma fractionation industry using Cohn cold ethanol fractionation outlined in *Table 2* generates as by-products of the main process substantial quantities of human protein precipitates that do not have any clinical value. The fraction 4 precipitates, which form part of the by-products, are rich in albumin and transferrin and, before recombinant proteins were available, were widely exploited as a source of insulin-like growth factors. However, the fraction 4 precipitate may also contain immunoglobulin and other high-molecular-weight components up to 25% of the total protein. It is also known to be contaminated with viruses present in the plasma fed into the fractionation process. Thus, although mammalian cells can be grown readily in medium supplemented with fraction 4 problems can arise from: (a) contamination of any cell product with redundant protein (especially in the case of monoclonal antibodies) or (b) infectious micro-organisms.

4.2 Subfractionation of Cohn fraction 4

To overcome the difficulties described above, procedures have been developed to reduce the amount of immunoglobulin and high-molecular-weight protein in fraction 4 derived cell culture medium supplement and to treat it to inactivate contaminating viruses. A suitable procedure is described in *Protocol 2*.

Only about 30% of the mass of a precipitated Cohn fraction is solid, the remainder being occluded liquor containing ethanol in the proportion used to produce the precipitate. Thus, when a Cohn fraction is collected it must be kept cold until it can be dispersed and the ethanol diluted. In particular, if the fraction is to be stored as a bulk paste it should be kept below −30°C. To make the precipitate workable it can be left overnight at 4°C, under these conditions some liquor will run out of the mass of precipitated protein, so it is important that the bulk fraction is left to thaw in an appropriate container.

Protocol 2. Subfractionation of Cohn fraction 4

1. Re-suspend the bulk fraction 4 precipitate in twice its weight of 0.9% (w/v) NaCl in water with gentle mixing.

2. Adjust the pH of the suspension to between 7 and 7.5 by addition of 1 M NaOH in water. Add the NaOH slowly with constant gentle stirring but thorough mixing to prevent prolonged exposure of the proteins in the fraction to localized high concentrations of NaOH. Continue stirring the mixture for 1 h after addition of the NaOH is completed. This and all subsequent steps may be performed at room temperature. The pH adjustment requires about 80 ml 1 M NaOH per 1 kg of original bulk precipitate but the precise quantity will depend on the severity of the centrifugation conditions under which the precipitate is collected.
3. Add 50% (w/v) polyethylene glycol (PEG), mean molecular weight 4000, in water to the re-dissolved fraction 4 protein solution to give a final PEG concentration of 10% (w/v). Add the PEG steadily with constant mixing and continue to mix for 30 min after the addition has been completed.
4. Centrifuge the suspension at 4000 *g* for 20 min at 20 °C. Decant and collect the supernatant; discard the precipitate.
5. Dilute the 10% PEG supernatant with three volumes of phosphate-buffered saline.
6. Add solid sorbitol to the diluted protein solution to give a concentration of 9% (w/v) and mix thoroughly until all of the sorbitol has dissolved.
7. Sterilize the protein solution immediately by filtration through a 0.2 μm pore size membrane.
8. Pasteurize the protein solution by heating at 60°C for 10 h.
9. Store the protein solution at 4°C.

The colour of the stabilized protein solution changes during the heating process from yellow to blue-green. This colour change can be prevented by the addition of ascorbic acid anti-oxidant but, as the product supports cell growth perfectly well and is stable for at least 6 months at 4°C without this addition, it is not normally included.

The use of 0.9% (w/v) NaCl in water to re-suspend the Cohn fraction precipitate and to dilute the supernatant to the PEG precipitate is most important. If the Cohn fraction precipitate is re-suspended in water, the subsequent addition of PEG will generate a very finely divided precipitate that cannot be collected by low-speed centrifugation. If the solution containing 10% (w/v) PEG is heated to inactivate infectious contaminants, the result is gross denaturation of the protein in the solution. To prevent this it is necessary to dilute the solution by the addition of three times its volume of phosphate buffered saline as described above. If water is used at this stage there is no immediately obvious effect on the product; however, it has been found that product in which water has been used for the final dilution is less stable than that in which phosphate buffered saline has been used. This instability results in loss of cell growth promoting activity and is most marked once the supplement has been incorporated into the final medium formulation.

The major components of Cohn fraction 4 and of the subfraction used as a medium supplement can be separated by polyacrylamide gel electrophoresis as shown in *Figure 1*.

5. The use of and adaptation of cells to growth in low-protein medium

A number of hybridoma cell lines have been grown successfully in medium supplemented with protein solution prepared from Cohn fraction 4 (4). The

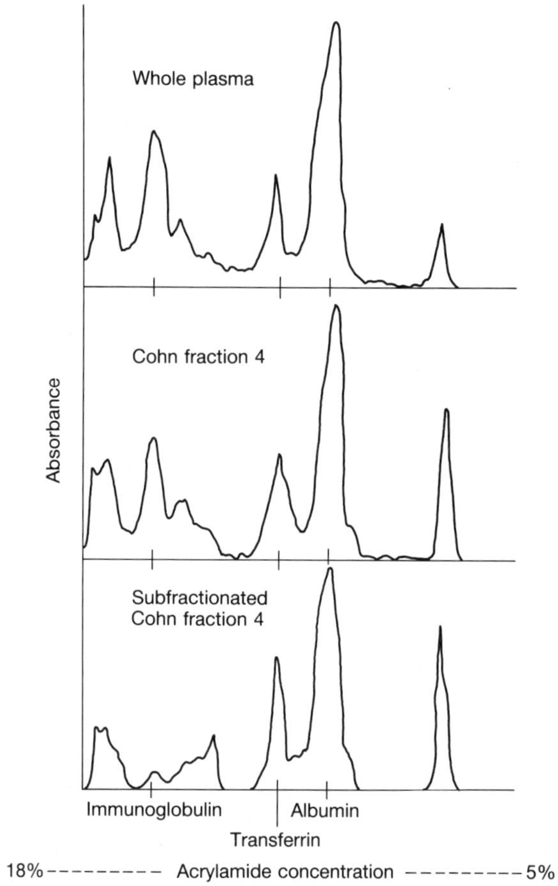

Figure 1. Analysis by polyacrylamide gradient gel electrophoresis of normal plasma, Cohn fraction 4, and subfractionated Cohn fraction 4.

formulation used has been 10% (v/v) fraction 4 solution in RPMI 1640 supplemented with 2 mM sodium pyruvate. Adherent MDCK cells have been grown on microcarriers in medium supplemented with only fraction 4 protein solution but their attachment and growth was improved by pre-treatment of the microcarriers with fibronectin (5).

All of the cell lines that have been grown in fraction 4 supplemented medium to date were originally established in fetal calf serum supplemented medium and the process of adaptation to the new formulation has been a critical part of the development (see also *Protocol 3*, Chapter 3). The first step in the adaptation is to establish the lowest concentration of fetal calf serum that can be used in the medium without adversely affecting the growth of the cells. This has been found to be at most 5% (v/v). The fraction 4 protein solution is then added to the medium at 10% (v/v) and the cells allowed to grow in the presence of both fetal calf serum and fraction 4 for at least 1 week. Throughout the adaptation process the cells are kept growing as rapidly as possible by frequent subculturing with a dilution of 1:4 in fresh medium. It is most important that the cells are not allowed to enter the late log or stationary phase of growth. The fetal calf serum content of the medium is then reduced in a series of steps to 3, 2, 1, 0.5, and 0% (v/v), the fraction 4 protein solution being kept at 10% (v/v). At each stage in the process the cells are taken through at least three or four passages and, if the growth rate of the cells is reduced, they are maintained at that stage until it recovers. When the cells are transferred from one fetal calf serum concentration to a lower one, cultures are maintained in both 'old' and 'new' formulations to provide some insurance against failure of the culture in the new conditions. In the case of a hybridoma the culture is also monitored throughout the adaptation process to ensure that antibody production is maintained.

The process of adaptation to low-protein medium is often performed in static cultures in tissue culture flasks but can be equally satisfactory in stirred cultures. Adaptation of cells in suspension culture has much to recommend it, if this is the type of system in which the cells will be used, as it reduces the problems that may be encountered when taking cells in low-protein medium from a static to a suspension culture. These problems can be minimized by seeding the suspension culture vessel at approximately 2×10^5 cells/ml, frequent subculturing to keep the cells in mid-log growth phase, and by keeping the stirrer speed fairly slow. Care must also be taken to ensure that the stirring is sufficiently vigorous to ensure that enough oxygen is adsorbed into the medium to meet the culture needs.

Once a population of cells has been adapted to growth in low-protein medium it is desirable to establish a cell bank in frozen storage. This can be done with cells adapted to low-protein medium by re-suspending packed cells from a late-log phase culture at 10^7 cells per ml in a mixture of 90% (v/v) human albumin solution (4.5 g/litre), 10% (v/v) DMSO. The cells are then aliquoted and frozen as normal (see Chapter 1).

Figure 2 shows the recovery of hybridoma cells adapted to growth in medium supplemented with subfractionated Cohn fraction 4 and then frozen. The recovery of the same adapted cells in medium containing fetal calf serum is also shown. These cells recovered from freezing and from transfer from stationary to stirred culture more rapidly in fetal calf serum supplemented medium but, ultimately, the cell growth rate was comparable in the two media. The specific antibody production rate was the same in the two media throughout the culture period.

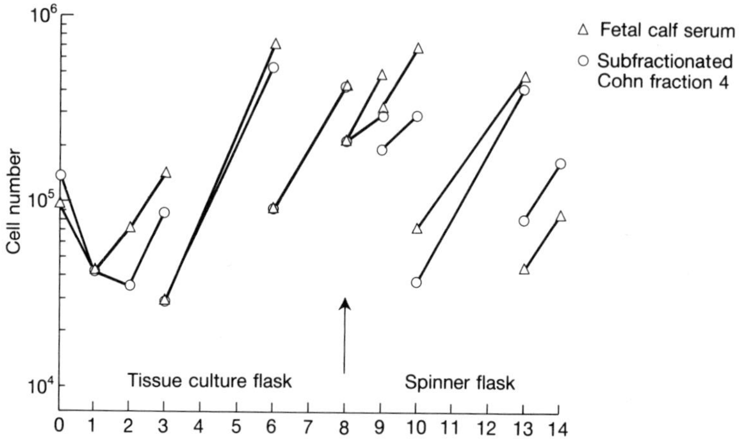

Figure 2. Growth of hybridoma cells from a frozen vial in RPMI 1640 medium supplemented with 10% (v/v) fetal calf serum or subfractionated Cohn fraction 4.

6. Process implications of the use of fractionated medium supplements

Media incorporating low-protein supplements retain much of the versatility of serum-supplemented media but offer a more standardized medium component than serum and easier product recovery from the lower protein background in the medium. However, there are complications introduced by the use of processed Cohn fractions. Cohn fraction 4 is collected at pH 5.2 and, although it is re-dissolved and processed, it is enriched in plasma components that are insoluble at that pH. Any subsequent downstream processing of the culture medium at approximately pH 5, (e.g. during ion-exchange chromatography) may be complicated by the development of an unwanted precipitate. This problem could be compounded if the culture supernatant has been concentrated by an ultrafiltration process because, although the polyethylene glycol used to precipitate the high-molecular-weight components of the fraction 4 solution has a molecular weight of 4000, it is a linear molecule and only passes very slowly through ultrafiltration membranes with pore sizes up to

30 000 nominal molecular weight limit. Thus the PEG in the culture supernatant at about 0.3% (w/v) may easily be concentrated tenfold during initial volume reduction by ultrafiltration and this will then be sufficient to aggravate severely the tendency of some components of the medium to precipitate.

7. General considerations

The use of human plasma fractions as the basis for preparation of an animal cell culture medium supplement has several points to commend it. The fractions are already produced routinely in bulk as part of the manufacture of therapeutic blood products and those that have been shown to be useful for cell culture are normally discarded. Consequently, their exploitation will not compromise established manufacturing operations and should not add greatly to costs.

The problems that arise from use of 'unrefined' plasma protein fractions are that they contain a substantial proportion of unwanted protein, especially immunoglobulin, and that they have to be regarded as being contaminated with infectious viruses. These problems have been addressed by the development of the subfractionation procedures described which both reduce the level of unwanted, interfering, proteins and make it possible to treat the product by pasteurization to inactivate viruses.

With particular reference to the production of therapeutic proteins from cell cultures there is probably considerable advantage to be gained from the use of human plasma proteins to supplement the medium. Experience with the established range of therapeutic blood products has demonstrated that redundant human plasma proteins in these products have a very low level of antigenicity. Thus any protein from the medium supplement left in the final product would not be expected to complicate its application by stimulating an immune reaction in the way, for instance, that bovine proteins would.

References

1. MacLeod, A. J. (1988). *Adv. biochem. Engng*, **37,** 41.
2. Kistler, P. and Friedli, H. (1980). In *Methods of Plasma Protein Fractionation* (ed. J. M. Curling), p. 3. Academic Press, London.
3. Hao, Y. L., Ingham, K. C., and Wickerhauser, M. (1980). In *Methods of Plasma Protein Fractionation* (ed. J. M. Curling), p. 57. Academic Press, London.
4. MacLeod, A. J. and Thomson, M. B. (1985). *Dev. Biol. Stand.* **60,** 55.
5. Sayer, T. E., Butler, M., and MacLeod, A. J. (1987). In *Modern Approaches to Animal Cell Technology* (ed. R. E. Spier and J. B. Griffiths), p. 264. Butterworths, Guildford.

3

Growth factors

NIGEL JENKINS

1. Introduction

The proliferation of mammalian cells is not only governed by the availability of simple nutrients (as found in microbial cultures), but also by the actions of peptide growth factors found in serum. These factors bind to specific receptors on the cell surface which stimulate the cell into a new round of mitosis. The importance of such events (often mediated via protein kinases) is underlined by the fact that many oncogenes in cancer cells are related to growth factors or their receptors. Several growth factors also play a role in controlling cell differentiation or in the uptake of essential nutrients.

In recent years there has been an explosion of information regarding mammalian growth factors and their mechanisms of action, which in some cases may prove confusing due to changes in terminology. In this chapter I will endeavour to provide a simple growth factor guide for the scientist wishing to construct mammalian cell cultures in a defined medium (i.e. without serum). There are advantages to be gained from culturing cells in a defined (serum-free) medium.

- The essential growth factor and other requirements can be defined for each cell line.
- There should be less batch variation in the components added to the medium, compared to serum.

The growth factor composition of sera may vary widely and, although some suppliers regularly test their serum batches for the ability to support cell growth, the efficacy of preparations tested on one cell line may not always apply to a different line with distinct growth factor requirements. On a small scale (<5 litre) the cost of constructing a serum-free medium may be identical to or even exceed that of serum-based media. It is only on a larger scale that the cost of defined media becomes significantly cheaper, particularly for cells with minimal growth factor requirements (see Section 4.1). Several companies now market serum substitutes for cell culture (e.g. Boehringer, Sigma); however, since the exact formulae of these preparations are not usually accessible, they are of little use in defining the growth factor requirements of cell lines.

I have outlined the structure and actions of the most common growth factors, and more comprehensive reviews on each factor are listed in the references. Further information on the biochemical properties of growth factors and their availability is contained in *Table 1*. The *specific activity of each growth factor preparation will vary between batches and with the source of the material*; therefore, only approximate dose ranges can be given. Storage and re-constitution protocols for these growth factors are given in Section 3.

The growth factor requirements for a wide range of mammalian cell types are listed in another volume of this series (1), and a general strategy for choosing media supplements is given in Section 4.1. It should be noted that: (a) transformed cell lines usually have fewer requirements for growth factors than primary cultures or cell lines with a finite lifespan; and (b) all cell lines diverge during extended periods of culture; therefore, *individual cell strains may differ in their growth factor requirements*.

In this chapter I have also examined carrier proteins, such as albumin and transferrin, which are not strictly speaking growth factors but are often included in defined medium formulations. Indeed, for some permanent cell lines these protein supplements are sufficient to maintain growth. Adherent cell lines may require extracellular matrix (ECM) proteins for the full effects of the growth factors to be expressed in serum-free medium, and I have included a short section on their properties.

Strategies for the adaptation of cells from serum-based to defined medium and the appraisal of their growth factor requirements are given in *Protocol 3*.

2. Growth factor properties

2.1 Fibroblast growth factors (FGF)

The principal mitogenic factors isolated from the bovine brain by heparin affinity chromatography can be resolved into two distinct peptides which share 55% homology (2): acidic FGF (pI 5.6) and basic FGF (pI > 9.0). Although both factors bind strongly to heparin (they are sometimes referred to as heparin-binding growth factors); only the potency of acidic FGF is augmented by heparin.

2.1.1 Acidic FGF (aFGF)

Also known as endothelial cell growth factor, this is the less potent of the two FGF mitogens (*Table 1*), even when augmented with 90 μg/ml heparin (which stabilizes aFGF tertiary structure and reduces its susceptibility to proteolysis). Variations in N-terminal processing during aFGF synthesis lead to the production of two structural variants: aFGF-I (134 amino acids) and aFGF-II (140 amino acids); these variants have similar biological potencies. The mitogenic action of aFGF on endothelial cells allows their propagation over 20–60 generations when grown on an ECM layer of fibronectin and collagen.

2.1.2 Basic FGF (bFGF)

The bFGF molecule also exists in several variants of similar biological potencies, which result from proteolytic processing (131 or 146 amino acids) or the use of alternative translational initiation sites (3). Basic FGF is produced by most cell types investigated *in vitro*, and the FGF receptor is ubiquitous. Thus, bFGF may act in an autocrine (stimulation of the same cell) or a paracrine fashion (one cell stimulating an adjacent cell). In ligand binding assays the FGF receptor binds *both* aFGF and bFGF (*Table 1*); however, this receptor has yet to be cloned and it is uncertain whether a single class of receptor is involved in mediating all the actions of FGF molecules. Nevertheless, if bFGF is included in a defined medium, it is unlikely that aFGF will also be required. Basic FGF elicits its mitogenic action through a calcium influx and activation of protein kinase C, and is a potent mitogen for many types of cultures derived from mesodermal cells, neuroectoderm, and several transformed cell lines (2). It also promotes the differentiation of adipocytes and ovarian granulosa cells, and induces the synthesis of ECM proteins such as collagen type IV. The bFGF itself may become incorporated into the ECM via its affinity for heparin, glycosaminoglycans, and collagen, and may only be released by enzymic hydrolysis of ECM proteins.

A bFGF supplement has been used to enhance the survival of new cell lines derived from primary cultures (4), and to eliminate the need for feeder cells during cloning of hybridomas. Feeder cells (e.g. irradiated fibroblasts or peritoneal exudate cells) are used to condition the medium with an undefined mixture of growth-promoting substances, particularly when the hybridomas are present at low densities (see Chapter 6).

2.2 Epidermal growth factor (EGF)

This mitogen has been isolated from mouse submaxillary glands and also human urine (formerly known as urogastrone). It is also available in recombinant form expressed in *E. coli* (*Table 1*). The mature peptide is only part of a much larger membrane-bound precursor and, in serum, circulates as a complex bound to several carrier proteins including a protease which cleaves the precursor (5). EGF is a potent mitogen for many primary cultures and epithelial and mesenchymal cell lines, and synergizes with other growth factors such as IGF-1 and TGFβ. Its membrane receptors have intrinsic tyrosine kinase activity and exist in two conformations: a high-affinity form (dissociation constant $K_d = 1\text{–}2 \times 10^{-10}$ M, representing 10% of the receptor population) and a low-affinity form ($K_d = 1\text{–}2 \times 10^{-9}$ M, 90% of the population). The significance of these two receptor populations is unknown at present.

2.3 Nerve growth factor (NGF)

NGF is purified from mouse submaxillary glands, bovine brain, and placenta.

Table 1. Properties of the common growth factors

Factor	Physical properties	Effective range	Receptor properties	Sources	Comments
Fibroblast growth factors (FGF)					
Acidic FGF (endothelial cell growth factor)	pI 5.6 Type I, 16.4 kd Type II, 17 kd	100–300 μg/ml (0.5–10 ng/ml with heparin)	Similar to basic FGF	Bovine brain[a–d]	Binds strongly to heparin
Basic FGF	Type I, 16.2 kd Type II, 18 kd	0.1–10 ng/ml	$K_d = 0.1–0.3$ nM $10^3–10^4$ receptors/cell	Bovine brain[a,c,d] Bovine recombinant[b] Human recombinant[b]	
Epidermal growth factor (EGF)	6 kd	1–20 ng/ml	$K_d = 0.1–2$ nM (2 classes) $1–4 \times 10^4$ receptors/cell	Mouse submaxillary gland[a–c] Human recombinant[b]	Carried as a 74 kd complex in serum
Nerve growth factor (NGF)	26.5 kd	5–10 ng/ml	$K_d = 0.01–1$ nM (2 classes) $1–5 \times 10^4$ receptors/cell	Mouse submaxillary gland[a–c]	Poor mitogen, but induces differentiation in neuroectoderm cells
Transforming growth factors (TGF)					
TGFα	7 kd	1–20 ng/ml	Binds to EGF receptors	Rat recombinant[c] Human recombinant[c]	Carried as a complex in serum
TFGβ$_1$	25 kd	0.1–5 ng/ml	K_d=25–40 pM $1–4 \times 10^4$ receptors/cell	Pig platelets (β_1 and β_2)[a,c,d] Human platelets β_1[c,d]	Synergistic with EGF or TGFα for growth in soft agar, but inhibits growth in monolayer cultures
TGFβ$_2$	25 kd				
Insulin	5.7 kd	1–10 μg/ml	$K_d = 1–5$ nM $10^3–10^4$ receptors/cell	Bovine pancreas[b]	Mitogenic and anabolic actions

Insulin-like growth factors (IGF)					
IGF-I (Somatomedin C)	7.6 kd	1–20 ng/ml	$K_d = 1–4$ nM 1–5 × 10^4 receptors/cell	Human recombinant[b,c]	Bind to carrier proteins in serum
IGF-II (Somatomedin A, MSA)	7.5 kd	1–5 ng/ml	$K_d = 10–20$ nM 5 × 10^5 receptors/cell	Rat liver[c]	
Platelet-derived growth factor (PDGF)	28–36 kd	1–20 ng/ml	$K_d = 20–200$ pM 1–2 × 10^5 receptors/cell	Pig platelets: BB dimer[b–d] Human platelets: AB dimer[a,c,d]	Binds to α_2-macroglobulin in serum
Interleukin-6 (IL-6)	20.6 kd	1–10 ng/ml	—	Human recombinant[b,c]	Can replace feeder layers for hybridomas
Albumin	66 kd	0.5–5 mg/ml	None	Bovine serum[a,b]	Lipid carrier
Transferrin	78 kd	1–10 μg/ml	$K_d = 2–20$ nM 1–4 × 10^4 receptor/cell	Human serum[a,c] Bovine serum[b]	Iron carrier
Matrix proteins					
Collagens	Multiple types	1–3 mg/ml for coating	None	Rat type I[a,b]	Maintains differentiated state
Fibronectin	440 kd	0.02 mg/ml for coating	$K_d = 1–2$ μM 2–6 × 10^3 receptors/cell	Human plasma[a,b]	Binds to cells and collagen
Laminin	400–1000 kd	0.02 mg/ml	multiple receptors	Mouse sarcoma[a]	—

[a] Sigma Chemical Company.
[b] ICN Biochemicals.
[c] Boehringer Chemical Company.
[d] British Biotechnology Ltd.

Like EGF it circulates in serum bound to carrier proteins including the protease kallikrein (6). It is commercially available as either the 140 kd serum complex (which is very stable but has low specific activity) or the 26.5 kd peptide dimer (which has a higher specific activity but is less stable). NGF is not a potent mitogen, but rather induces the differentiation and enhances the survival of sympathetic neurons and PC12 phaeochromocytoma cells in culture. NGF receptors are found on several neuroectoderm tissues, in both high-affinity ($K_d = 10^{-11}$ M) and low-affinity conformations ($K_d = 10^{-9}$ M) as shown in *Table 1*. Unlike the EGF receptor, the recently cloned NGF receptor does not display tyrosine kinase activity (6).

2.4 Transforming growth factors (TGF)

First discovered in the medium of retrovirus-transformed rat kidney fibroblasts, these factors have since been identified as natural secretion products of several non-transformed cell lines (7). The ability of TGF to cause anchorage-independent growth in soft agar has been resolved into two fractions (TGFα and TGFβ) which are distinct and unrelated molecules binding to different receptors on the cell surface.

2.4.1 TGFα

This factor bears 30% homology to EGF, and indeed binds to EGF receptors with equal potency. Its actions are similar to those of EGF (although it is more potent in promoting the vascularization of tissues, ref. 8). Therefore, if EGF is included in a defined medium, then TGFα is unlikely to be required. Also, some cell lines (e.g. A431) which produce endogenous TGFα may actually be inhibited from growing by EGF (9). TGFα is synthesized as a large membrane-bound precursor which is subsequently cleaved by proteases into the mature peptide. TGFα is found naturally in the embryonic kidney, adult brain, pituitary gland, skin, and placenta. It is available as a recombinant product of either the rat or human gene expressed in *E. coli* (*Table 1*).

2.4.2 TGFβ

This factor is commonly purified from human platelets (which predominantly secrete TGFβ-1), although other species such as pig also secrete a related peptide (TGFβ-2) which has 63% homology. Both variants of TGFβ bind to the same receptor and have equal biological potencies in most assays (*Table 1*). They are part of a larger gene family which also includes peptides controlling reproductive functions (e.g. inhibin). The ubiquitous TGFβ receptor is not a tyrosine kinase, but specific enhancer regions have been found in the promoter regions of several genes which respond to TGFβ.

The biologically inactive TGFβ precursor (220–235 kd) circulates in blood as a complex bound to carrier proteins and possibly a protease. Since most mammalian cells in culture secrete small amounts of TGFβ precursor and have TFGβ receptors, conversion of the inactive complex to the active dimer must be a crucial step in the regulation of TGFβ activity. Once released, the TGFβ dimer has only a short half-life, even though it binds α_2-macroglobulin in serum.

TGFβ promotes the growth of several mesenchymal cell lines in soft agar (e.g. fibroblasts) in combination with either TGFα or EGF. In contrast, TGFβ alone *inhibits* the growth of many cell lines in monolayer culture, which may result from its ability to stimulate the secretion of ECM proteins (collagens, fibronectin, and glycosaminoglycans) and protease inhibitors (10). Monolayer growth inhibition by TGFβ is most prominent in epithelial cells, endothelial cells (antagonized by bFGF), stem cells, and lymphocytes.

2.5 Insulin

Several of the mitogenic actions previously ascribed to insulin may be due to insulin-like growth factor contamination of the original preparations. Insulin is quite unstable at 37 °C (particularly in media with high levels of cysteine) with over 90% of its activity being destroyed over 1 h (11). It is often added to media at relatively high concentrations (*Table 1*) and, because insulin has a weak affinity for IGF-1 receptors, it may activate the same mitogenic responses as IGF-1 at these high doses. However, insulin is also added to most serum-free media for its ability to promote anabolic metabolism (e.g. glucose uptake and oxidation, synthesis of glycogen, amino acid transport). The ubiquitous insulin receptor is a dimer which has tyrosine kinase activity (12).

Insulin has traditionally been purified from pig or bovine pancreas but, more recently, human recombinant insulin (expressed in *E. coli*) has become available (*Table 1*). Care should be taken to re-constitute insulin in acid solutions (in which it is most stable) and to ensure adequate zinc is present in the medium since this element is essential for the biological activity of insulin (see *Protocol 2C*).

2.6 Insulin-like growth factors (IGF)

These peptides have evolved by gene duplication and are homologous to insulin (13). They were first discovered as the serum components which mediate the effects of growth hormone on cartilage—hence their former name of somatomedins. IGFs are bound to carrier proteins in blood which play an important role in regulating their delivery to tissues (14). Many of the metabolic actions of IGF overlap with insulin; however, they are more potent mitogens (*Table 1*) and synergize with FGF and PDGF.

2.6.1 IGF-I

Formerly known as Somatomedin-C, this peptide has been purified from liver and is now available in human recombinant form (*Table 1*). IGF-1 is secreted in low amounts by most tissues, and acts on a wide range of mesenchymal cells. Although IGF-I binds to the insulin receptor with low affinity, its main actions are mediated by a specific IGF-I receptor which has intrinsic tyrosine kinase activity. This receptor also binds IGF-II and, to a lesser extent, insulin (14).

2.6.2 IGF-II

Formerly known as multiplication-stimulating activity (MSA), this factor is produced by a more restricted range of tissues (e.g. fetal liver, muscle, skin, and adult brain). The IGF-II receptor is also less common and does not bind significant amounts of either insulin or IGF-I. This receptor is found on adipocytes and (in contrast to insulin and IGF-I receptors) does not display tyrosine kinase activity.

2.7 Platelet-derived growth factor (PDGF)

This factor is a common component of serum (40–60 ng/ml) and its peptide structure varies depending on the species of platelets from which it was purified (15). Pig platelets produce a BB peptide dimer which is the most stable form and has the highest affinity for PDGF receptors (the BB dimer is also available in recombinant form, *Table 1*). Human and bovine platelets secrete predominantly the less active AB dimer (the A chain is a 60% homologous peptide). In serum both forms of PDGF bind to α_2-macroglobulin. The name PDGF is somewhat misleading since it is also secreted by endothelial and placental cells, and by several SV40-transformed lines such as BHK and 3T3 fibroblasts (16). PDGF synergizes with EGF and IGF-I in stimulating normal Balb/C 3T3 fibroblasts to proliferate. Its action is restricted to mesenchymal and neuroectoderm cells which have specific PDGF receptors exhibiting tyrosine kinase activity (12). In addition to its potent mitogenic effects, PDGF also inhibits the differentiation of myoblasts and is chemotactic for monocytes, neutrophils, and fibroblasts.

2.8 Interleukin-6 (IL-6)

A complete description of all the interleukins and haemopoietic growth factors is beyond the scope of this chapter, and the reader is referred to a recent review on this topic (17). However IL-6 (formerly known as interferon-β_2) deserves a special mention due to its mitogenic action on hybridomas and stem cells (18). IL-6 can replace the use of feeder cells (such as peritoneal exudate cells, spleen cells, fibroblasts, or thymocytes) during cloning by limiting dilution and in the post-fusion stages of hybridoma culture. It is secreted by monocytes, T helper lymphocytes, and endothelial cells and is now available in recombinant form (*Table 1*). Mouse and human IL-6 are equally effective in stimulating hybridomas created from most species, including difficult lines such as human–mouse and rat–mouse hybrids. IL-6 also synergizes with other interleukins to enhance antibody production.

2.9 Carrier proteins

For the full effects of growth factors to be demonstrated, other supplements need to be added to most serum-free media formulations. Such factors serve to support the transfer of lipids (albumin) or trace elements (transferrin) to the cell (19).

2.9.1 Albumin and lipids

Conventional albumin preparations (prepared by cold ethanol Cohn fractionation or by heat shock) have varying amounts of lipid contamination. These lipids are predominantly fatty acids which may be beneficial to the cell, since free fatty acids are not only difficult to solubilize but may also be toxic. We have found that CHO cells do *not* grow well in a serum-free medium where de-lipidated albumin was provided and no additional lipids were supplemented. Lipids can therefore be provided by following one of these regimes:

- albumin preparations containing lipids (e.g. BSA from Cohn fraction 5 at 0.5–5 mg/ml, Sigma A4503)
- fatty-acid free albumin (Boehringer 1081489, 0.5–1 mg/ml) mixed with soya bean lipids (Boehringer 652229, 20–100 μg/ml); see reference 11 for protocol
- fatty-acid free albumin (Boehringer 1081489, 0.5–1 mg/ml) plus a lipoprotein supplement (e.g. Excyte, Miles Biochemicals)

Albumin also acts as a buffering and detoxifying agent (by binding endotoxins and pyrogens), as a carrier for trace elements, vitamins, and hormones (e.g. steroids, thyroxine), and serves to protect cells from shear forces.

2.9.2 Transferrin

Transferrin is the major iron carrier protein of vertebrate serum. Most mammalian cells have specific transferrin receptors which take up the transferrin/Fe^{3+} complex as their major source of iron, an essential trace element. There is evidence that transferrin also has growth factor properties in addition to its iron transport role, and it binds trace elements such as vanadium. In cell culture medium (pH 7.4) iron salts will dissociate to form the oxidized ferric iron (Fe^{3+}) which is only sparingly soluble without transferrin. Human transferrin binds tightly to cells from a wider range of species compared to bovine transferrin (20), and is therefore preferred in defined media formulations. It is commercially available either as a 30% iron-saturated complex (Boehringer) or iron-free (*Table 1*). When using the iron-free form, ensure that sufficient iron salts are added to 30% saturate the transferrin (*Protocol 1*).

Protocol 1. Iron saturation of transferrin

- Tissue-culture grade water (Sigma W3500) or water purified using a cartridge filtration system (e.g. Milli-Q, Millipore Inc)

Protocol 1. *continued*

- Analar grade HCl
- Analar grade ferric citrate ($FeCl_3 \cdot 6H_2O$). *Warning*: this compound is hygroscopic and should be stored in a vacuum desiccator
- Sodium salt of Hepes (Sigma H0763)
- Iron-free transferrin (see *Table 1* for suppliers)

Method

1. Prepare 500 ml of a 10 mM HCl solution in water.
2. Dissolve 1.35 g $FeCl_3 \cdot 6H_2O$ into the HCl solution (this compound is hygroscopic, therefore weigh it quickly). Use large volumes to minimize weighing errors.
3. Prepare a 25 mM Hepes buffer solution in water.
4. Reconstitute lyophilized human transferrin at 30 mg/ml in the Hepes buffer. Check pH is 7.4, adjust with 2 M NaOH or 2 M HCl if required.
5. Immediately add the 100 μl of the $FeCl_3$ solution to 3.9 ml of the transferrin solution to achieve an optimal 30–40% iron saturation.
6. Filter sterilize (e.g. Gelman 0.22 μm Acrodiscs); store at 4°C in sterile Eppendorf tubes for up to 6 months.
7. Vortex thoroughly before use. Discard vials which are still cloudy.

2.10 Extracellular matrix proteins

Some adherent cells do not bind directly on to plastic and require extracellular matrix (ECM) proteins for efficient plating and growth (21). Some ECM proteins (e.g. fibronectin) appear in serum, and are secreted by cells. Others such as the collagens are secreted by cells either constitutively or in response to growth factor stimulation (bFGF, TGFβ). These growth factors may become embedded in the matrix.

Collagen type I is available commercially, prepared from rat tendons (*Table 1*). It is stored in an acidic aqueous solution (0.1 M acetic acid), and is used to coat dishes at 1–3 mg/ml overnight. It is important to wash the collagen film on the dish surface with Hepes buffer (25 mM) or cell culture medium (pH 7.4) before use. The film can be sterilized using ultraviolet irradiation, or by a quick ethanol rinse.

Fibronectin is purified by heparin affinity chromatography from human plasma (*Table 1*) and can be stored at 4°C. It binds to collagens and to a specific cellular receptor (integrin) which is found particularly on mesenchymal cells such as fibroblasts. Before use, ensure the protein is in solution by vortexing vigorously, and plate on to culture dishes as a film of 0.02 mg/ml in sterile PBS for 2 h at 37°C. Wash dishes with sterile medium before use. For cells which are particularly difficult to plate in serum-free medium, a pre-coat of sterile poly-D-lysine (Boehringer, final concentration 0.5–1 mg/ml in sterile PBS) may be applied to the dishes (30 min at 37°C) before coating with fibronectin (22).

Laminin is another ECM protein found in basement membranes (*Table 1*) and is sometimes used as an alternative to fibronectin, especially for plating epithelial cells (coat at 0.02 mg/ml, 2 h, 37°C).

3. Re-constitution and storage

Growth factors are generally small peptides (5–30 kd) which have a tendency to adhere to plastic or glass surfaces at low concentrations (1–1000 ng/ml). Most growth factors are supplied in a lyophilized state, and it is advisable to store them lyophilized (at the manufacturer's recommended temperature) until required for the first cell culture. It is also desirable to reconstitute and dilute lyophilized growth factor preparations in a 1 mg/ml solution of high-quality carrier protein (e.g. protease-free albumin, Sigma A3294), and (where factors need frozen storage) to prepare each factor in aliquots small enough to avoid repeated freeze–thawing. We prefer to snap-freeze these preparations by immersion in liquid nitrogen (*CAUTION*: insulated gloves and a face mask must be worn when handling liquid nitrogen), after which they may be stored in an ordinary freezer (−20°C). Add the growth factor supplement to the medium *immediately* before treating the cells. It is also worth remembering that in the uncomplexed state most growth factors are small enough to pass through a standard dialysis membrane.

Protocol 2. Storage and re-constitution of growth factors

- tissue-culture grade water (see *Protocol 1*)
- phosphate-buffered saline tablets (PBS, Sigma P4417, for *Protocol 2A*)
- analar HCl (for *Protocol 2B*)
- protease-free bovine serum albumin (BSA, Sigma A3294)
- sterile filtration cartridges (Gelman Acrodisc, 6224192)
- sterile pipette tips and plastic storage tubes (Eppendorf), and sterile 22G needles
- a source of liquid nitrogen and a Thermos flask.

A. *Growth factors stable at neutral pH (EGF, aFGF, bFGF, IL-6)*

1. Make a 1 mg/ml solution of BSA in 0.15 M PBS, sterile filter into a sterile universal tube through a 0.22 μm filter. Store at 4°C for a maximum of 2 months.
2. When required for the first cell culture, reconstitute the lyophilized growth

Protocol 2. *continued*

factor at 100–500 × the final concentration required for culture (*Table 1*) in the PBS–BSA solution at room temperature.

3. Sterile filter (0.22 μm pore size) and dispense into sterile Eppendorf tubes using sterile pipette tips.

4. Snap-freeze tubes by immersion in liquid nitrogen for 1 min, store at a minimum temperature of −20 °C.

5. To thaw, puncture top of the tube with a sterile 22G needle (to allow escape of air) and immerse the tube in a 37 °C water bath. Remember to swab the outside of the tubes in 70% ethanol: 30% water before opening.

6. Dilute the growth factor to the required final concentration (*Table 1*) in fresh medium at 37 °C and use the medium immediately.

B. *Growth factors stable at acidic pH (IGF-1, TGFα, PDGF, and TGFβ)*

1. Make a solution of 4 mM HCl in tissue-culture grade water.

2. Dissolve the protease-free BSA at 1 mg/ml. When required for the first cell culture, reconstitute the lyophilized growth factor at 100–500 × the final concentration required for culture (*Table 1*) in this HCl–BSA solution at room temperature.

3. Follow step **3** from *Protocol 2A*. For IGF-1 and TGFα snap-freeze and store frozen (steps **4–6** of *Protocol 2A*). PDGF and TGFβ are more stable when stored at 4 °C (then follow step **6** of *Protocol 2A*).

C. *Growth factors requiring metal ions (insulin, NGF, transferrin)*

1. *Insulin*. Follow *Protocol 2B* but, where the zinc content of the manufacturer's preparation is less than 0.3%, add zinc sulphate heptahydrate (Sigma Z0251) to the HCl–BSA solution at 4.48 μg/ml before adding the insulin (step **2** of *Protocol 2B*).

2. *NGF*. Follow *Protocol 2A* but, where the zinc content of the manufacturer's preparation is less than 0.3%, add zinc sulphate heptahydrate (Sigma Z0251) to the PBS–BSA solution at 4.48 μg/ml before adding the NGF (step **1** of *Protocol 2A*).

3. *Transferrin*. For preparations already saturated with iron (e.g. from Boehringer) follow *Protocol 2A*. For iron-free transferrin (e.g. from Sigma) follow *Protocol 1*.

---

4. Adaptation of cells to defined medium

4.1 Choice of supplements

The most difficult part of the adaptation process is to predict exactly *which* supplements will be required for each cell line in a defined medium. However, some guidance can be given from work on similar or identical cell lines, as outlined in previous chapters on this subject (1, 19). The following description is given as a rough guide only—as stated in the introduction individual lines and even laboratory strains of cells can differ markedly in their growth factor requirements.

As a general principle cells fall into four categories in terms of growth factor requirements.

(a) *Transformed cell lines* which pass through the G1/G0 phase of the cell cycle with requirements for maintenance factors only. Examples of this type of cell are CHO lines, SV40-transformed BHK cells, melanomas, mouse hybridomas, and myelomas (23). Although these cells do express basic FGF receptors, their requirement for this factor is not absolute. Their needs can generally be met by bovine insulin (5 μg/ml), together with the carrier proteins BSA (1–5 mg/ml Cohn fraction 5, for lipids and vitamins), human transferrin (5 μg/ml for iron delivery), and trace element supplements (*Table 2*) which may be deficient in some basal media formulations.

(b) *Mesenchymal cells* are likely to require maintenance factors as detailed in Section 4.1 (a) in addition to growth factors. Cells in this category include fibroblast-derived lines (e.g. Balb/C 3T3, Swiss 3T3), adipocytes (which also respond to IGF-II), endothelial cells, smooth muscle cells, and neuroectoderm cells (e.g. glial cells). To minimize the number of growth factor tested, it should be recalled that:

   i EGF and TGFα share the same receptor (EGF is more commonly available)

   ii acidic FGF and basic FGF share the same receptor (basic FGF is more potent)

   iii high doses (>3 μg/ml) of insulin purified from mammalian sources may contain sufficient amounts of IGF-1 to support growth

Table 2. Trace element and amino acid supplements added to RPMI-1640 for growth of CHO cells in serum-free medium (all reagents are Sigma cell culture grade)

L-Alanine	8.9 mg/l
L-Glutamine	294 mg/l
Putrescine	0.16 mg/l
Sodium pyruvate	110 mg/l
Ferrous sulphate	0.83 mg/l
Zinc sulphate	0.86 mg/l
Copper sulphate	2.5 μg/l
Sodium selenite	1.73 μg/l

All the above are made up in Milli-Q water at 100× final concentration and filtered through a 0.2 μm filter before use.

iv the BB dimer of PDGF (from porcine platelets) is more potent than the AB dimer found in human platelets

v TGFβ is growth *inhibitory* to most cell lines in monolayer cultures, and only causes anchorage-independent growth in soft agar in combination with either TGFα or EGF

As an example, the mouse embryo fibroblast (NIH-3T3) cell line can be grown in a serum-free medium (24) consisting of Ham's F12/DMEM (Gibco) supplemented with transferrin (25 μg/ml), insulin (10 μg/ml), EGF (100 ng/ml), bFGF (100 ng/ml), and PDGF (0.5 U/ml) provided the dishes are pre-coated with fibronectin and poly-D-lysine (see Section 2.10).

(c) *Epithelial cells* are generally the most difficult to grow in a defined medium, particularly to retain their differentiated phenotype. Maintenance factors are usually required (as detailed in Section 4.1 (a)), and EGF (5–10 ng/ml) is added to most cell cultures. Extracellular matrix proteins such as collagen, fibronectin, or laminin are often required to promote cell adhesion. Also non-peptide factors such as retinol (50 ng/ml) and corticosteroids (10–100 ng/ml) may be required. Epithelial cells which have been grown in defined media include those isolated from the mammary gland, testis, ovary, pituitary gland, prostate, and lens tissue, and the reader is referred to references 1 and 11 for complete media formulations.

(d) Certain cell types may also require cell-specific factors such as NGF (for neuroectoderm, and PC12 cells), IL-6 (for hybridomas at clonal dilutions), and growth hormone (for hepatoma cells).

Protocol 3. Adaptation of cells to defined medium

The main purpose of this protocol is to enable cells to adapt at a pace which should not compromise cell viability or product synthesis.

1. Ensure that the cells are growing well in their present, serum-based medium. Determine the optimal seeding density which will allow 5–20 fold growth over the period of the experiment (do not seed at very low densities).
2. If base medium is of the type developed for serum-free or low-serum cultivation (e.g. RPMI 1640, Ham's F12, Iscoves medium, or Ham's MCDB series) proceed to stage 3. Otherwise, replace the base medium with one or more of these formulations (maintaining the same serum concentration) and evaluate the growth of cells over 5–8 days. Choose the base medium which best supports cell growth.
3. Passage the cells (using trypsin–EDTA for adherent cells, (see Chapter 1) into a 24-well plate and culture with a gradation of serum concentrations for 3–4 days. For example, if the original culture was in 10% serum, set up

cultures in quadruplicate wells at 10 (control), 8, 6, 4, 2, and 1% serum. Add the full supplements chosen for the defined medium (Section 4.1) from this stage onwards. Also ensure that any selective agents required to maintain copy number in recombinant cell lines (e.g. methotrexate for $DHFR^+$ selection) are included in the medium.

4. Choose the serum concentration which supports cell growth to at least 70% of the control wells. Maintain cells at this concentration until they achieve 70% confluence (or for non-adherent cells $5–7 \times 10^5$/ml). If none of the cultures have reached this state by 7 days, consider modifying the supplements (Section 4.1). Passage into a 25 cm^2 flask, then a 75 cm^2 flask to expand the cultures. If required, test for secretion of the desired cell product at this stage (there should be no loss of production).
5. It is prudent to make frozen stocks of cells at $4–6 \times 10^6$/ml (Chapter 1) from cultures growing well at different stages in the adaptation protocol, as a safeguard against cells failing to grow at a lower serum level. Once the stage of <0.5% serum has been reached, use a sterile freezing mixture of base medium with 25 mg/ml BSA and 10% DMSO.
6. Set up another 24-well plate at a lower gradation of serum concentrations. For example, if 4% serum supports growth at stage 4, choose 4 (control), 3, 2, 1, 0.8, 0.4, 0.2% serum. Grow cells for 3–4 days.
7. Repeat stages **4** and **5** until serum is eliminated.
8. Titrate growth factors (*Table 1*) to eliminate those which are unnecessary, and to reduce the doses used to a minimum.
9. For permanent cell lines, clone by limiting dilution (at an average of 1–2 cells per well) in 96-well plates and select clones with best growth properties and/or product secretion. Note that irradiated feeder layers may be required at low cell densities (see Section 2.1.2).

An example of the time taken for a recombinant CHO cell line to adapt to a serum-free medium (Section 4.1, point (a)) using this protocol is shown in *Figure 1*.

Acknowledgements

This work was supported by the SERC Biotechnology Directorate and a consortium of the following companies: Celltech, Glaxo, Imperial Chemical Industries, Porton International, Smith–Kline Beecham, and Wellcome Biotechnology.

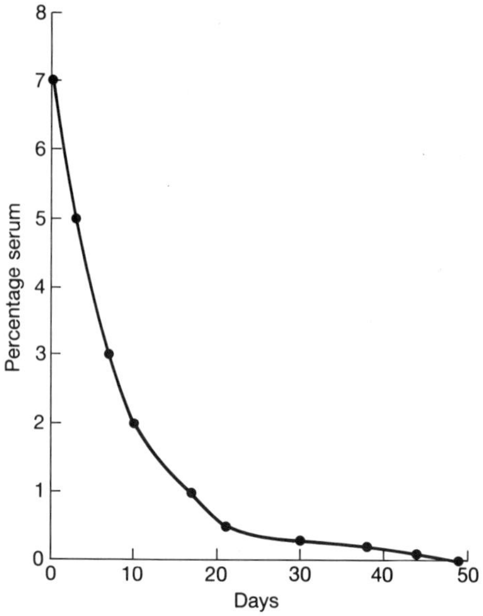

Figure 1. Time course of the adaptation of a recombinant CHO cell line to serum-free medium. Cells were originally grown in RPMI 1640 containing 7% adult bovine serum and were adapted to the same medium containing maintenance factors only (Section 4.1, point (a)).

References

1. Maurer, H. R. (1986). In *Animal Cell Culture, A Practical Approach* (ed. R. I. Freshney), p. 13. IRL Press, Oxford.
2. Burgess, W. H. and Maciag, T. (1989). *Ann. Rev. Biochem.* **58,** 575.
3. Rifkin, D. B. and Moscatelli, D. (1989). *J. Cell Biol.* **109,** 1.
4. Gospodarowicz, D., Ferrara, N., Schweigerer, L., and Neufeld, G. (1987). *Endocrinol. Rev.* **8,** 95.
5. Carpenter, G. (1985). *J. Cell Sci. Suppl.* **3,** 1.
6. Hempstead, B. L. and Chao, M. S. (1989). In *Recent Progress in Hormone Research*, Vol. 45 (ed. J. H. Clark), p. 441. Academic Press, London.
7. Sporn, M. B., Roberts, A., Wakefield, L. M., and Crombrughe, B. (1987. *J. Cell Biol.* **105,** 1039.
8. Derynck, R. (1988). *Cell* **54,** 593.
9. Parkinson, E. K. and Yeudall, W. A. (1991). In *Culture of Specialized Cells: Epithelial Cells*, Vol. 1 (ed. R. I. Freshney). Wiley-Liss. (In press.)
10. Moses, H. L., Coffey, R. J., Leof, E. B., Lyons, R. M., and Kesi-Oja, J. (1987). *J. Cell Physiol. Suppl.* **5,** 1.
11. Barnes, D., and Sato, G. (1980). *Anal. Biochem.* **102,** 255.
12. Williams, L. T. (1989). *Science* **243,** 1564.
13. Daughaday, W. H. and Rotwein, P. (1989). *Endocrinol. Rev.* **10,** 68.

14. Baxter, R. C. (1988). *Comp. Biochem. Physiol.* **91B,** 229.
15. Deuel, T. F., Silverman, N. J., and Kawahara, S. (1988). *BioFactors* **1,** 213.
16. Pledger, W. J., Estes, J. E., Howe, P. H., and Leof, E. B. (1984). In *Mammalian Cell Culture. The Use of Serum-Free Hormone-Supplemented Media* (ed. J. P. Mather), p. 1. Plenum Press, New York.
17. Nicos, N. A. (1987). *J. Cell Physiol. Suppl.* **5,** 9.
18. Sugasarawa, R. J. (1988). *Biotechnology* **6,** 895.
19. Ham, R. G. and McKeehan, W. L. (1979). In *Methods in Enzymology*, Vol. 58 (ed. W. B. Jakoby and I. H. Pastan), p. 44. Academic Press, London.
20. Young, S. P. and Garner, C. (1990). *Biochem. J.* **265,** 587.
21. Kleinman, H. K., Luckenbill-Edds, L., Cannon, F. W., and Sephel, G. C. (1987). *Anal. Biochem.* **166,** 1.
22. McKeehan, W. L. and Ham, R. G. (1976). *J. Cell Biol.* **71,** 727.
23. Schacter, E. (1989). *Trends Biotechnol.* **7,** 248.
24. Greenwood, D., Srinivasan, A., McGoogan, S., and Pipas, J. M. (1989). In *Cell Growth and Division: A Practical Approach* (ed. R. Bersaga), p. 37. IRL Press, Oxford.

4

Genetic engineering of animal cells

CAROLINE MACDONALD

1. Introduction

The commercial exploitation of mammalian cells has largely been for the production either of viruses for vaccines, or for monoclonal antibodies. This situation is changing, however, as the techniques for introducing and expressing foreign genes in animal cells become more routine. This chapter will review the most commonly used methods for introducing genes into animal cells and compare their advantages and disadvantages. It will also describe the strategies for constructing expression vectors and for selecting cell lines expressing heterologous genes at a high level.

2. Transfer of nucleic acid into animal cells

The role of DNA in the transfer of properties between prokaryotes was first shown in 1944 when Avery and colleagues demonstrated that bacterial transformation was mediated by DNA transfer (1). It was not until 1962, however, that the biochemical transformation of thymidine kinase deficient (TK^-) mouse cell mutants to a TK^+ phenotype was reported by Szbalska and Szbalski (2). They added fragments of herpes simplex virus (HSV) DNA to cells and showed that it was possible to isolate cells which could grow in HAT-selective medium (containing hypoxanthine, aminopterin, and thymidine) due to the presence of the purine and pyrimidine salvage pathway enzymes, thymidine kinase (TK) and hypoxanthine–guanine phosphoribosyl transferase (HGPRT). However, they were unable to demonstrate that the cells had actually taken up the gene from the HSV DNA fragments, and the possibility that some of the cells had reverted to the TK^+ phenotype could not be discounted.

In 1968 Burnett and Harrington reported the transformation of cultured cells by treatment with DEAE–dextran and incubation with viral DNA (3). However, the method which has been most widely used, based on co-precipitation of the DNA with calcium phosphate, was first described for the infection of cells with adenovirus 5 DNA by Graham and van der Eb (4), and for the transfer of cellular genes by Wigler and co-workers (5). A variety of modifications to these original methods have been described, and a range of alternative techniques for introducing genes into cells also exist and will be described below. These methods fall into two groups—those designed for transient gene expression, and those in which stably transformed cell lines are isolated.

In transient expression systems DNA is introduced into the cells and, after 1 to 3 days, the cells are harvested and the products of the expression of the recombinant gene analysed. The level of expression must be high, and a large proportion of the cells must take up and express the gene for sufficient product to be made. In stable transformation the efficiency of delivery is less important because only a small proportion of the cells which take up DNA will become permanently transformed. It is therefore essential to include a selectable marker within the DNA, so that the cells which are expressing the DNA can be identified and selected. Stable transformants must have either integrated the DNA into the chromosome, or the recombinant DNA must be capable of efficient, autonomous replication so that the foreign sequences are not diluted out by cell division. Some vectors and some gene transfer methods are particularly suitable for stable transformation, and some are more appropriate for transient expression.

Many of the strategies for transferring genes into animal cells involve the introduction of purified plasmid DNA. It is important that this DNA is in good condition and free from contaminating RNA and bacterial chromosomal DNA. A number of methods exist for preparing pure plasmid DNA, including lysis of the bacteria with alkaline sodium dodecyl sulphate (SDS), followed by purification of the plasmid DNA from the bacterial DNA by banding on a caesium chloride gradient. *Protocol 1* has given excellent results in our laboratory; however, good results have also been obtained using commercially available rapid purification systems, e.g. Qiagen Hi Purity (Diagen). Some variability between DNA preparations has been observed with the 'kit' methods—possibly due to contamination with different amounts of bacterial DNA, the presence of which has been reported to interfere with transfection. A final caesium chloride purification step is therefore recommended if the DNA is to be expressed in mammalian cells.

Protocol 1. The purification of DNA

1. Set up a 3 ml culture from an inoculum established from a single bacterial colony containing plasmid DNA and grow overnight at 37°C.
2. Inoculate this 3 ml into a 1-litre culture of L-broth[a] and incubate overnight at 37°C with shaking.
3. Pellet the bacteria by centrifugation at 5000 r.p.m. for 10 min at 4°C.
4. Wash the bacterial pellet in 20 ml cold TGE (25 mM Tris/10 mM EDTA/

50 mM glucose), pH 8 and centrifuge again. All subsequent lysis procedures should be carried out on ice.

5. Re-suspend the bacteria in 20 ml TGE containing 10 mg/ml lysozyme which has been freshly made up. Incubate on ice for 5 min.
6. Add 40 ml freshly made alkaline SDS[b], invert bottle to mix, and incubate for a further 5 min.
7. Add 30 ml 3 M sodium acetate (pH to 4.8 with glacial acetic acid) and mix, gently at first and finally vortexing to complete the mixing and precipitation of the bacterial DNA.
8. Incubate on ice for 5 min then centrifuge at 5000 r.p.m. for 10 min.
9. Transfer the clear supernatant to four 50 ml centrifuge tubes and extract with an equal volume (~ 25 ml) of phenol:chloroform[c].
10. Centrifuge at 5000 r.p.m. for 10 min and collect the top, aqueous phases. Combine in two fresh tubes, add 25 ml propan-2-ol to each, vortex, and leave for 5 min.
11. Centrifuge again, combine the pellets, and dissolve in 10 ml 0.1 M sodium acetate (pH to 6.0 with glacial acetic acid). Add 25 ml of cold (−20°C) ethanol and precipitate at −20°C for 1 h.
12. Pellet the nucleic acid by centrifugation at 5000 r.p.m. for 10 min, wash with ethanol, and dry in a vacuum oven for 1 h.
13. Dissolve the pellet in 10 ml TE (10 mM Tris-HCl pH 7.6/1 mM EDTA) and treat with 0.1 mg/ml RNase A for 30 min at 37°C.
14. Add 1 ml 2.5 M NaCl, extract with phenol/chloroform, precipitate with 2.5 vols ethanol at −20°C, vacuum dry, and dissolve in 1 ml TE.
15. Purify the plasmid DNA by caesium chloride density gradient centrifugation. Weigh the DNA solution, add an equal weight of CsCl, mix gently to dissolve, and add 1/10 volume 10 mg/ml ethidium bromide[d].
16. Centrifuge in a vertical or angle rotor for 2.5–4 h at 90 000 r.p.m. or 18–24 h at 50 000 r.p.m. or equivalent.
17. Following centrifugation, clamp the tube and illuminate with long-wave UV light to visualize the bands of DNA. Locate the lower band of plasmid DNA, puncture the tube by inserting an 18 gauge needle just below the band, and collect the DNA.
18. Remove the ethidium bromide by extracting five times with an equal volume of pentanol discarding the upper phase each time.
19. Dialyse overnight against TE to remove the pentanol and caesium chloride.
20. Calculate the concentration of the DNA by measuring absorption at 260 nm and assuming that the absorbance of a 50 μg/ml solution in a 1 cm

Protocol 1. *continued*

quartz cuvette is 1. The purity of the DNA is calculated from the ratio of absorbances at 260/280 nm which should fall between 1.65 and 1.85.

[a] L broth: dissolve 10 g bactotryptone, 5 g yeast extract and 5 g NaCl in 950 ml of deionized water, adjust pH to 7.0 with 5 M NaOH, make up to 1 litre, autoclave, and store at room temperature. Just before use add antibiotics if required and glucose to a final concentration of 1%.

[b] Alkaline SDS should be made up just before use by adding 1 ml 2 M NaOH, 1 ml of 10% SDS, and 8 ml water.

[c] Phenol/chloroform is a 1:1 mixture of phenol and 'chloroform'. Phenol contains impurities which should be removed by redistillation. 'Chloroform' is a 24:1 mixture of chloroform and isoamyl alcohol.

[d] Ethidium bromide is made up in 10 mM Tris-HCl, pH 7.5. Detailed methods for the preparation and handling of reagents and buffers for molecular biology are given by Sambrook *et al.* (6).

The main methods used for DNA transfer are described in the following subsections and summarized in *Table 1*.

Table 1. Methods for introducing DNA into cells

Method	**Advantages**	**Disadvantages**	**Reference number**
DEAE–dextran	Simple	Transient expression only	3
Calcium phosphate	Simple	Unsuitable for suspension cells	4, 7–14
Liposomes	Simple	Relatively unproven	15
Micro-injection	Efficient	Technically difficult	16–19
Electroporation	Good for non-adherent cells	No co-transfection	20
Protoplast fusion	Good for non-adherent cells	Variable results	21
Retrovirus infection	Efficient	Cell type restricted by tropism; low coding capacity	22

2.1 Calcium phosphate co-precipitation

This method is also known as DNA transformation (by analogy with bacterial transformation) or transfection (because originally viral genes were transferred). It involves incubating cells with DNA in the form of a calcium phosphate co-precipitate. The precipitate sediments on to the cells, becomes adsorbed on to the cell membrane, and is then taken up into the cytoplasm through a calcium-requiring process. After a period of incubation, which is normally 4–16 h, the precipitate is washed off from the cells. Calcium phosphate can be replaced with strontium phosphate for introducing genes into cells whose growth is inhibited or which are lysed in the presence of calcium (7). Cells which are able to take up DNA in this way can take up large amounts of DNA including unlinked marker genes (8). This phenomenon, known as co-transformation, can be used to select for expression of DNA which does not contain a selectable gene, by forming a mixed co-precipitate with DNA which includes a selectable marker. If an excess of the first DNA is used then almost all of the cells expressing the selectable marker will also be expressing the first DNA sequence.

A detailed discussion of procedures for calcium phosphate precipitation has been published by Gorman (91); but an abbreviated, standard protocol, *Protocol 2*, follows.

Protocol 2. The introduction of DNA into cells using calcium phosphate precipitation

1. Set up cells in 60 mm dishes in 5 ml culture medium at 50% confluence. Incubate overnight.
2. Three hours prior to transfection change the culture medium and add gentamycin (0.25 mg/5 ml) as a precaution against bacterial contamination from the DNA preparation.
3. Prepare precipitate.
   (a) Mix together 20 μg DNA, 60 μl 2.5 M $CaCl_2$ (filter sterilized), and sterile distilled water to a final volume of 500 μl.
   (b) Add 500 ml 2 × HBS[a] to a sterile polypropylene tube.
   (c) Add the DNA mixture to the 2 × HBS dropwise while vortexing. Alternatively, air can be added to the tube as a stream of bubbles from an electric pipetting device while the DNA is being dropped in.
4. Allow precipitate to stand for 15 to 30 min then add 1 ml precipitate to each 60 mm dish. Incubate for 4 to 24 h at 37°C.
5. Aspirate medium and rinse twice to remove DNA from cells. Incubate in fresh medium containing gentamycin.

[a] 10× HBS (Hepes buffered saline) contains 8.18% NaCl (w/v), 5.94% Hepes (w/v), and 0.2% $NaHPO_4$ (w/v). Adjust pH to 7.12, filter sterilize, and dilute to 2× before use.

A variety of modifications to *Protocol 2* have been described. These include the addition of chloroquine at a concentration of 200 μg/ml to the DNA solution (10). DEAE–dextran, first used to increase the efficiency of virus infection, has also been used to improve the uptake of plasmid DNA (11) for transient expression, but this method cannot be used to make permanent transformed cell lines (9). A dimethyl sulphoxide (DMSO) boost (12) or glycerol shock (13), added immediately after removal of the DNA, will improve the transfection efficiency in many cells. Overnight treatment of cells with 2–10 mM butyrate following glycerol shock can also increase the efficiency of transient expression (14).

2.2 Liposome encapsulation

DNA can be encapsulated in liposome vesicles and then delivered to the cells by polyethlene glycol-mediated cell fusion (*Protocol 3*). Recently, a commercially-available cationic liposome-mediated method has been described. Unilamellar liposomes are formed from a positively-charged lipid named DOTMA (N-[1-(2,3 dioleyloxy)propyl]-N,N,N-trimethyl ammonium chloride) and complexed with DNA or RNA. These complexes are readily taken up by cells, and impressive expression levels have been des cribed for a range of different cell types using a very straightforward procedure (15).

Protocol 3. The introduction of DNA into cells using DOTMA reagent

1. Plate the cells in 60 mm dishes in 5 ml Dulbecco's modification of Eagle's medium (DMEM) or similar growth medium supplemented with 10% fetal calf serum (FCS) and incubate until around 80% confluent. For each 60 mm dish use the quantities given in the following steps.
2. Dilute 10 μg DNA to 50 μl in water.
3. Dilute 40 μg DOTMA reagent (obtainable from Life Technologies as Lipofectin, or from Boehringer Mannheim) to 50 μl in water. (The DNA and DOTMA reagent must be diluted separately in water in order to prevent precipitation.)
4. Combine the DNA and DOTMA in a polystyrene snap cap tube, mix gently without vortexing, and allow to stand at room temperature for 15 min. (*Note*: Polypropylene tubes should not be used as the mixture tends to adhere.)
5. Wash the cells twice in serum-free medium then incubate in 3 ml serum-free medium. The presence of serum will inhibit the uptake of DNA.
6. Add 10 μl DOTMA–DNA complex to the cells dropwise, dispersing the complex as evenly as possible over the cells.
7. Incubate the cells for 5 to 24 h at 37°C; then change to medium supplemented with serum.

2.3 Micro-injection

Small volumes (10–100 fl) of DNA can be micro-injected into tissue culture cells using glass capillary micropipettes operated by gentle air pressure provided by a syringe and injecting into either the cytoplasm or directly into the nucleus. Nuclear micro-injection results in very high frequencies (up to 20%) of stable integration of the desired sequence (16). An alternative method, micro-injection by iontophoresis rather than pressure, gives no net fluid displacement and so minimizes shearing and allows very large amounts of DNA to be injected (17), although more DNA is degraded prior to integration in this method. It has been reported that iontophoretic injection gives multiple integration without the trandem insertions which are commonly found in pressure-injected cells. A simpler method has also been described (18) in which the cells are incubated in a DNA–saline solution and then pricked over the nucleus with a microneedle, but only about 2% of the cells incorporate DNA in a stable manner. Another variation, known as 'laser-aided' transfection, exists in which the cell membrane is perforated with a finely focused laser beam (19). However, all of these techniques require expensive and sophisticated equipment and a considerable degree of operator expertise and are beyond the scope of this chapter.

2.4 Electroporation

The cell membrane can also be permeabilized by exposing the cells to a high-voltage electrical impulse *Protocol 4*; (reference 20). DNA present in the cell suspension can be taken up, presumably through micropores which are formed transiently in the cell membrane. The efficiency of uptake by electroporation can be very high and, in general, cells contain only one copy of the foreign DNA. This is presumably because the DNA is present at low concentration in the solution and, therefore, it is unlikely that more than one, or at most a few, copies will enter the cells. This is in contrast to the situation which occurs with protoplast fusion and transfection, and it means that two or more genes can be transferred together only if they are physically linked. The conditions for electroporation vary considerably with different cell lines and preliminary experiments must be carried out to determine the optimum conditions. For this reason, although electroporation works well with many cells which do not take up DNA by calcium phosphate precipitation, e.g. myeloma and hybridoma cells, considerable optimization may be needed to get conditions for efficient transfer. A variety of different equipment can be used for electroporation ranging from a simple home-made pulse generator modified from an electrophoresis power pack to specialized electroporation pulse generators. Most of the low-cost generators produce exponentially decaying impulses from a bank of capacitors, whereas the more expensive, specialized instruments produce square or nearly square waves.

The efficiency of electroporation is influenced by the following parameters.

(a) *The strength of the electric field.* The maximum voltage which can be generated by the instrument may influence its suitability. If the voltage is too low, the membrane may not permit transfer of DNA. If it is too high, irreversible damage to the cells may occur. If the electroporation is carried out in phosphate-buffered saline or serum-free medium using a volume of 1 ml, then ideally the generator should be able to deliver a voltage of at least 1500 V because the optimum voltage required for any particular cell type can vary from 250 to 1250 V/cm.

(b) *The shape and duration of the generated electric field*. Some instruments allow the operator to choose the length and shape of the electric pulse. It is not clear, however, that square waves produce significantly better results. A single electric pulse of 20–100 msec duration is normally used.

(c) *The reproducibility of the impulse*. Considerable variation can occur between experiments and this can be partly attributed to the generation of irreproducible electric fields.

(d) *Temperature*. Although it has been proposed that better results should be obtained at 4°C rather than room temperature due to enhanced cell viability, in practice absolute efficiency appears to be better at room temperature.

(e) *Electroporation medium*. The electric field strength is affected by the ionic composition and conductivity of the medium. In general, good results are obtained with Hepes or phosphate-buffered saline (HBS or PBS, respectively) however, for cells with poor viability in these buffers it may be necessary to use culture medium for the electroporation.

(f) *Preparation of DNA*. Electroporation has been observed with DNA at concentrations from 1 μg/ml to 1 mg/ml, ranging in size from small plasmids (5 kb) to fragments, cosmids, and minichromosomes of 30 to 150 kb. Higher levels of both transient and stable transformation have been reported when linear DNA is used in preference to supercoiled circular plasmid DNA.

(g) *Cells*. As in all gene transfer techniques the condition of the cells is crucial. Optimum transfer occurs with cells which are in mid-log phase, suspended at a concentration of 5×10^6 to 10^7 cells/ml. Cells which grow in suspension often require significantly greater field strengths than do adherent cells; voltages up to 4 kV/cm have been reported.

Protocol 4. The introduction of DNA by electroporation

1. Harvest the cells in mid-log phase of growth.
2. Wash twice in Hepes buffered saline (HBS).
3. Re-suspend in HBS at 10^7 cells/ml.
4. Add the DNA in 10 mM Tris-HCl, pH 7.6, 1 mM EDTA to the cells to a final concentration of 10–100 μg/ml. Transfer the suspension to a sterile electroporation chamber.
5. Pulse according to the instructions of the instrument manufacturer, following the pulse on an oscilloscope if possible to ensure reproducibility.

6. Incubate the mixture at room temperature for 10 min.
7. Plate out the cells in medium containing serum.
8. Replace the medium after 24 h.

2.5 Protoplast/spheroplast fusion

DNA can be introduced into mammalian cells by direct fusion of the cells to bacterial protoplasts (21). The bacterial cell wall is removed by lysozyme digestion; then the bacteria and mammalian cells are fused together using polyethylene glycol as fusogen (*Protocol 5*). This method has the advantage of not requiring a DNA purification step, but is limited to DNA which has been cloned into bacterial vectors. It is important tht the lysozyme digestion of the bacterial wall is completed but, even so, it is anathema to many cell geneticists to add bacteria intentionally into a culture of mammalian cells!

Protocol 5. Protoplast/spheroplast fusion

1. Prepare a 10 ml culture of bacteria.
2. Pellet bacteria by centrifugation at 3000 r.p.m.
3. Re-suspend pellet in 1 ml 25% sucrose/50 mM Tris-HCl, pH 8.0, in a polypropylene tube.
4. Centrifuge at 13 000 r.p.m. for 1 min and re-suspend cells in 200 μl of 20 mg/ml lysozyme (made up freshly) in 25% sucrose/50 mM Tris.
5. Incubate on ice for 10 to 20 min.
6. Add 50 μl 0.5 M EDTA, pH 8.0, and incubate on ice for a further 10 min.
7. Add 1 ml serum-free medium slowly.
8. Centrifuge at 13 000 r.p.m. for 1 min.
9. Re-suspend bacterial protoplasts in 1 ml serum-free medium and store on ice.
10. Prepare cells using standard trypsinization procedures. Centrifuge and re-suspend the cells in serum-free medium.
11. Mix together 10^9 bacterial and 5×10^6 mammalian cells and centrifuge at 2000 r.p.m. Re-suspend the pellet in 200 μl 46% polyethylene glycol (w/v) in serum-free medium[a] and mix together for 1 min. Add 1 ml serum-free medium dropwise; then add a further 10 ml serum-free medium.
12. Centrifuge, re-suspend the cells in medium containing serum, and plate out into dishes.

[a] Autoclave 4.6 g PEG 1500 and, while still molten, make up to 10 ml with serum-free medium. Roll to mix. Use within 1 week.

2.6 Retroviral infection

The most efficient method of introducing genes into cells is to use a recombinant viral vector (22). Viruses have evolved effective systems for infecting cells, based on binding to cell-surface receptors, but, in general, cells are lysed as a result of viral infection. Retroviruses are an exception to this rule, and retroviral vectors offer an efficient gene delivery method which is particularly useful in situations where the number of available recipient cells is limited, e.g. for introducing genes into embryos.

In contrast to most other viruses, retroviruses do not replicate episomally within the host cell but become incorporated into the host cell genome. Integration proceeds via a specific, controlled mechanism and results in the incorporation of a single copy of the viral genome. Most retroviral vectors are derived from either avian retroviruses, such as Rous sarcoma virus (RSV), or murine retroviruses such as Moloney murine leukaemia virus (MoMLV) but contain only the long terminal repeat (LTR) sequences from the virus and the packaging signal.

The genome of an infectious retrovirus contains three structural genes, *gag, pol,* and *env*, which code for the viral core proteins, reverse transcriptase, and the envelope proteins, respectively. This coding sequence is flanked by LTR sequences at either end. In addition, the virus contains primer binding sites for reverse transcription of viral RNA adjacent to the LTRs; a packaging sequence ψ, which is necessary for efficient packaging of the viral RNA into virions; and viral splice donor and acceptor sequences which are used for the production of a subgenomic mRNA. The strategy for vector design is to replace the *gag, pol*, and *env* genes to generate replication-defective vectors which require complementation of viral functions *in trans*. The only sequences which need to be retained on the vector are the two LTRs which are needed for integration, the adjacent primer binding sites, and the packaging signal. The remaining viral genes are provided either by replication in the presence of helper virus, or, more frequently, by propagating the vector in a packaging cell line which provides the missing functions. The packaging cell line has been genetically engineered to contain a defective retroviral genome integrated within the cellular DNA in order to provide the viral functions *in trans*. The integrated genome lacks a packaging signal and is not, therefore, packaged to give virus particles. The virus secreted by the packaging cell line will contain the vector sequence which can infect target cells at high efficiency. The host range of the infectious virus is determined by the envelope gene contained within the packaging cell line. Thus to produce vector derived from MoMLV which will infect mouse cells the vector should be propagated in an ecotropic packaging cell line such as ψ_2 (23) or ψ_{cre} (24), whereas, if the target cells are human, feline, or canine, an amphotropic cell line such as PA317 (25) or ψ_{crip} (24) must be used.

The stages involved in constructing a usable retroviral expression system follow.

(a) The gene of interest is cloned into a plasmid which contains the relevant viral sequences, i.e. the LTRs and the packaging signal.
(b) The vector is introduced into a packaging cell line by transfection or electroporation.
(c) Clones of transfected cells are selected on the basis of expression of an antibiotic resistance gene (*Protocol 6*).
(d) The packaging cell lines are grown up, frozen stocks prepared, and the secreted virus titred (*Protocol 7*) and assayed for the presence of wild-type recombinants (*Protocol 8*).
(e) Viral stocks are harvested from appropriate packaging lines and used to infect target cells.

Stocks of packaging cell lines and target cells must be assayed for the presence of wild-type virus at regular intervals in order to ensure that recombination between vector and integrated virus, or vector and naturally-occurring endogenous retroviruses has not occurred.

Protocol 6. Selection for cells which are expressing *neo*

G418 titration

1. Set up cells at an appropriate density: for most cell lines this will be around 8×10^3 cells/cm^2 if adherent, or 10^5 cells/ml if non-adherent. However, very rapidly growing cells may require to be seeded at a lower density and it may be possible to seed very small cells at a higher density. Incubate cultures overnight to allow the cells to attach.
2. Add G418 to the culture medium at concentrations ranging from 250 to 1500 μg/ml. Incubate for 3 days.
3. Remove spent medium and add fresh medium containing G418. Incubate for a further 3 days.
4. Change medium again and incubate for a further 3 days.
5. Scan the culture for the presence of living cells by inverted phase microscopy.
6. Choose the lowest concentration of drug which results in complete cell death for the selection of transfectants.

(*Note*: Drug concentrations can be made up more accurately if a high concentration of G418 in medium is diluted with medium to provide the lower concentrations. The activity of G418/mg of dry weight varies from batch to batch and the percentage of active material of a given lot should be noted.)

Protocol 6. *continued*

Selection of transfectants

1. Set up cells at an appropriate density[a]. Introduce DNA using one of the protocols described above.
2. Add G418 24 h after removal of the DNA. Incubate the cells.
3. Replenish the medium with G418 every 3–4 days until colonies of transfected cells are visible and can be picked. Transfer colonies of cells into individual wells of a 24-well plate by gently touching the cells with the end of a disposable Pasteur pipette[b]. Alternatively, cloning rings may be used if preferred.

   If it is important that all transfected cell lines are independent in origin, remove only one colony, plus a 'back-up' for use if the original fails to transfer, from each of the original containers. If large numbers of independent clones are required, consider using 24-well plates for the transfection, or transfecting in suspension and seeding out immediately afterwards into 24-well plates.
4. Expand the cells in the absence of G418 until the population is large enough for freezing and for testing for drug resistance.

[a] The density must be such that the cells can undergo a further 3–4 divisions, because G418 is only toxic to dividing cells. An alternative procedure to the one given is to seed the cells at one-third confluence, add DNA, and then divide the cells by one-third to one-fifth as soon as the DNA is removed.

[b] Disposable Pasteur pipettes give superior results to glass pipettes because the cells cling better to the rough edges of the plastic.

Protocol 7. Virus titration

Vectors which contain a selectable marker, such as *neo*, can be conveniently titred by infecting a target cell line and counting the number of *neo*-resistant colonies.

1. Set up a culture of the packaging cell line in a 25 cm^2 flask and grow until approximately one-third confluent, such that the cells can continue to grow in log phase for a further 24 h.
2. Remove medium from the flask, add fresh medium, and incubate at 37°C for 24 h.
3. Remove the culture medium and filter through a 0.45 μm cellulose acetate filter (ICN–Flow) to remove cellular debris.
4. Make serial dilutions of the medium in fresh medium by stepwise dilution to 10^{-1}, 10^{-2}, 10^{-3}, 10^{-4}, and 10^{-5}.
5. Set up target cells, e.g. 3T3 cells, 24 hours prior to use at 3×10^5 cells per 25 cm^2 flask in 5 ml medium and incubate overnight.

6. Remove medium from the target cells and add 2 ml of virus-containing medium plus 8 μg/ml Polybrene (Aldrich) to promote virus binding.
7. Incubate for 2 h, add a further 3 ml of fresh medium (without removal of the original virus-containing medium), and incubate overnight.
8. Remove medium from the cells, trypsinize, and re-seed in a 75 cm^2 flask in 10 ml medium and incubate overnight.
9. Add G418 at a concentration (determined previously by titration) which will kill all uninfected cells (*Protocol 6*).
10. Change medium, replenishing the G418, every 3–4 days until colonies are visible.
11. Count colonies at all appropriate dilutions of virus-containing medium and calculate the virus titre[a] in colony-forming units per ml (cfu/ml).

[a] The yield of virus from low-producing cell lines can be increased by harvesting the virus after 48 h rather than 24 h (at step **2**) or by ultracentrifugation to concentrate the virus.

An estimate of the titre of stocks of virus which do not contain a selectable marker can be made by dot blot hybridization to a viral probe. However, reverse transcriptase (RT) assays used by some workers do not distinguish between packaged vector and 'empty' virions secreted by the packaging cell lines since these particles contain RT although they lack RNA.

Protocol 8. Assay for recombinant helper virus

The presence of recombinant helper virus from ecotropic packaging cells is detected by the XC plaque assay, modified from Rowe (26). Amphotropic virus is not detected by this assay because it does not produce syncytia on XC cells, and a cell line such as S^+L^- (sarcoma-positive, leukaemia-negative) mouse cells (27) must be used instead.

1. Inoculate 3T3 cells at 3.5×10^5 cells/dish in a 5 cm diameter tissue culture dish in 5 ml medium and incubate overnight.
2. Harvest the virus from the infected cells to be tested as described in *Protocol 7*.
3. Remove the 3T3 culture medium and replace with a 0.5 ml aliquot of the virus-containing medium to be tested plus 0.5 ml of fresh medium containing 20 μg/ml Polybrene.
4. Incubate the cells for 2 h, add a further 3 ml of fresh medium, and incubate the cells overnight.
5. Culture the cells for another 4 days and then expose to a lethal dose of UV irradiation (1600 erg/mm^2). This is done by aspirating the medium, re-

Protocol 8. *continued*

moving the lids from the dishes, and exposing the cells to an unfiltered, bactericidal 254 nm UV light (a distance of 8 in for 1 min).

6. Immediately after irradiation add 10^6 XC cells to the 3T3 cells in 5 ml medium and incubate for 2 days.
7. Change the medium and incubate the cells for a further 2 days.
8. Fix the cells and stain with methylene blue–basic fuchsin stain (3.3 g of methylene blue (Sigma) and 1.1 g basic fuchsin (Sigma) per litre of methanol).
9. Examine the stained plates for the presence of syncytia by inverted microscopy.

An alternative procedure for detecting helper virus in target cells is to assay for the presence of virus containing the *neo* gene by marker rescue. This will detect recombination which results in neo^+/replication-competent virus, but may not detect all possible helper virus recombinants. Cells should be passaged for 2 weeks to allow potential helper virus to spread until all of the cells in the culture are infected. The medium is then harvested from the cells and assayed for the ability to form G418-resistance colonies on naïve cells.

The procedure for infecting adherent target cells is identical to that given for infecting 3T3 cells for assaying the virus titre, except that the virus stock is not diluted prior to use. The infection of non-adherent cells is similar, but the cells should be centrifuged and re-suspended in fresh medium plus virus; for example, 1 ml of virus supernatant plus 1 ml of fresh medium could be used to infect 5×10^5 cells. In addition, the cells must be centrifuged and re-suspended in fresh medium at each medium change. The efficiency of infection should be high, provided a high-titre virus stock is used, and it may be possible to obtain infected target cells in the absence of selection by screening a number of clones. If the culture is infected at a multiplicity of infection (m.o.i.) of one, then, statistically, the expectation is that 63% of the cells in the population will be initially infected.

3. Viral expression vectors

The first mammalian expression vectors (*Table 2*) were genetically engineered viruses capable of multiplication in the host cells and propagation as infectious virions. This strategy meant that there was a limitation on the size of the sequence incorporated into these vectors due to the need to package the viral genome into the virion. Thus, the inserted sequence usually replaced essential viral genes which had to be provided in some other way. This could be either by propagating the vector as a mixed stock with a helper virus or by providing the missing genes in the host cell line (i.e. a helper cell line).

Table 2. Mammalian expression vectors

Vector	Strengths	Limitations
Transient vectors		
SV40 replacement	High yield	Limited host range, small capacity
SV40 *ori*$^+$/COS	High yield; no virus produced	Limited to COS cells
Polyoma	High yield	Limited host cells; small capacity
Vaccinia	Large capacity; potential for vaccine	Cloning difficult
Adenovirus	Wide host range	Cloning difficult due to size; transient
Herpes	Broad host range	Cloning difficult
Baculovirus	Inexpensive, safe, high yields	Incorrect glycosylation
Stable vectors		
BPV	High copy number	Restricted host range
EBV	Large capacity	Cloning difficult; cannot be used in rodent cells
BK		Not widely available
AAV	Inducible; transient, or stable	Small capacity; need helper virus
Sinbis	Short-term	Produce virus
Retrovirus	Efficient	Single copy; small coding capacity; relatively unstable

3.1 Transient vectors

3.1.1 SV40 vectors

The first mammalian vectors were replacement vectors based on the monkey virus SV40. SV40 is a small DNA virus with a circular genome of 5224 bp which has an early region which contains two genes (coding for two transforming proteins known as large T and small t antigen); and a late region which contains three structural genes. The expression vectors developed from SV40 fell into two types: those in which the structural genes had been deleted, i.e. late region replacement vectors; and those in which the T antigen coding region had been removed, the early region replacement vectors.

These vectors allow DNA to be introduced into the cells by infection, a process which is approximately 100-fold more efficient than DNA-mediated transformation. However, they have a number of disadvantages, including a limited host range (SV40 will only replicate in monkey cells) and the ability to accommodate only a small insert, and are, therefore, mainly of historical interest. A more versatile transient expression system based on SV40 has been developed by Gluzman (28) using COS cells. This cell line was derived from the monkey line CV-1 by the incorporation of an origin-deleted SV40 genome into the host cell chromosome. These cells express T antigen but are unable to replicate the SV40 DNA due to the deletion in the origin. However, plasmid DNA which contains a sequence of approximately 85 bp which includes the viral origin of replication can be replicated to a high copy number ($2–4 \times 10^5$ copies/cell) (29). This system is useful for testing a vector for good levels of expression but cannot be used to generate stable cell lines.

3.1.2 Polyomavirus vectors

Polyoma is a papovavirus which is closely related to SV40 and has a similar genomic structure. The viruses differ in their host range: polyoma replicates in mouse cells, rather than monkey cells as is the case for SV40. A polyomavirus helper cell line (MOP cell line) can provide polyoma early region genes to complement functions missing from polyomavirus vectors (30). However, these vectors are rarely used because they share the disadvantages of SV40 vectors, namely, restricted host range, small capacity, and suitability only for transient expression.

3.1.3 Vaccinia virus vectors

Vaccinia virus is a large DNA virus with a genome of approximately 187 kb which can be genetically modified and used to express foreign genes for vaccination. The virus can accommodate multiple inserts while still retaining infectivity for a wide range of species and cell types. However, the large size of the genome makes cloning difficult and the foreign gene(s) must be inserted into the virus by homologous recombination. This is done by constructing a plasmid vector with the foreign gene expressed from a vaccinia promoter and flanked by non-essential vaccinia sequences, e.g. the thymidine kinase (*tk*) gene. The plasmid DNA is introduced into cells infected with vaccinia and recombination between the *tk* sequences in the plasmid and in the viral genome results in the insertion of the foreign gene into the *tk* gene of the virus. These recombinant viruses occur at low frequency but can be selected for by plaque assay on tk^- cells in the presence of 5-bromodeoxyuridine (BUdR). Wild-type virus will be killed because thymidine kinase will metabolize the BUdR and it will be incorporated into nucleic acid, whereas recombinant virus will fail to metabolize the BUdR and will therefore survive (31).

Vaccinia virus vectors expressing immunogenic proteins will grow in culture, and experimental animals infected with these recombinant vaccines have been shown to be protected against the infectious agent. The stability and ease of administration of the vaccine means that recombinant vaccinia vectors expressing the protective antigens for combinations of pathogenic agents offer considerable promise for use as a polyvalent live vaccine (32).

3.1.4 Adenovirus vectors

Adenoviruses, in particular the human Ad2 and Ad5 serotypes, have been used for the development of vectors which are capable of efficient infection and have a broad host range. Sequences of up to 7.5 kb can be inserted into vectors by cloning into the E3 region which is not essential for *in vitro* growth. Other vectors, for example, those in which the E1 region is replaced, can be propagated in a helper cell line (the 293 cell line) which contains the E1 region of Ad5 (33).

Adenovirus vectors can express high levels of a foreign protein because transcription from the major late promoter is very efficient. Host protein synthesis is shut-off very efficiently which facilitates product recovery although it results in cell death (34). Adenovirus vectors may be useful as vaccines because they have been used to express genes such as the hepatitis B virus surface antigen, and they will replicate in cell lines such as Vero and WI-38 which are accredited for vaccine production (35).

3.1.5 Herpesvirus vectors

Herpes simplex virus (HSV) vectors have been used for the transient expression of genes such as hepatitis B virus S gene and a chimeric chicken ovalbumin gene. HSV vectors have a broad cell type and species host range as well as a large coding capacity (36). Herpesviruses are able to remain latent within an infected cell which may enable cell lines which are stable producers of recombinant proteins to be isolated using herpesvirus vectors. The cell line could be propagated with the vector in a latent form, and then induced to produce the foreign protein at an appropriate time. This approach has been successfully used to express bovine growth hormone from a herpesvirus saimiri vector in a marmoset T cell line (37).

3.1.6 Baculovirus vectors

Expression vectors have been developed from baculoviruses such as *Autographa californica* nuclear polyhedrosis virus (AcNPV) which infects a number of insect cell lines including the fall armyworm cell line Sf9. The polyhedrin gene, a non-essential gene expressed at high level, is replaced by the foreign gene which is expressed from the polyhedrin promoter (38). Baculovirus vectors can be used to produce high levels of protein in cells grown in fermenters at lower temperatures and in less expensive medium than that used for mammalian cell culture. However, there are differences in the glycosylation patterns observed in proteins produced in insect cells and this may make this system unsuitable for the synthesis of human proteins for therapeutic use.

3.2 Stable expression vectors

The initial aproach to high-level, stable expression in mammalian cells was to use as vectors viruses which would replicate episomally in the host cell to high copy number without causing cell lysis. Vectors derived from bovine papillomavirus have been used extensively in industrial laboratories for engineering cell lines for the large-scale production of therapeutic proteins.

3.2.1 Bovine papillomavirus vectors

Rodent cells can be transformed by bovine papillomavirus (BPV) and the viral genome is generally maintained as an episome which will replicate to about 100 copies per cell. Not all of the viral genome is necessary for transformation, which can also occur when only 69% of the viral genome is present, and so the remaining coding capacity of the virus can be used for cloning. Stable expression of rat pre-proinsulin has been achieved by introducing the BPV subgenomic transforming fragment BPV_{69T} linked to the pre-proinsulin gene into cells by calcium phosphate transfection and identifying the transformed clones on the basis of morphology (39).

BPV vectors have two disadvantages the first being a tendency to integrate into the host cell genome which means that the advantage of episomal replication is lost. Second, the vector will only replicate in mouse cells thus limiting the range of cells in which the recombinant gene can be expressed.

3.2.2 Epstein–Barr virus vectors

Epstein–Barr virus (EBV) vectors have been developed to overcome the limitation of host cell species of BPV vectors. EBV can infect human B lymphocytes and some epithelial cells *in vivo*, and will replicate as an autonomous episomal element. The *cis*- and *trans*-acting viral elements which are required for episomal replication are known, and it has been shown that vectors carrying only the *cis*-acting element, *ori*P, can replicate in EBV-transformed human B lymphoblasts (40). These cells act as helper cells, like COS cells for SV40 replication, supplying the gene which encodes EBV nuclear antigen (*EBNA-1*) which is essential for replication. Vectors which contain both *ori*P and *EBNA-1* will replicate in cell lines such as HeLa which lack EBV sequences, but the copy number is much lower than that observed with cells which contain EBV (41).

EBV vectors can replicate in EBV-transformed cell lines without causing cell death, and can be maintained by selection at a copy number of up to 50 plasmids per cell. In addition to having a broader host range than BPV (although they are unable to replicate in rodent cells), EBV vectors are usually unrearranged in recipient cells and appear to have a very low frequency of mutation.

3.2.3 BK virus vectors

Milanesi *et al.* (42) have described a vector derived from the human papovavirus, BK virus which persists episomally in human cells. They have used this vector to express the thymidine kinase (*tk*) gene from herpes simplex virus type 1 (HSV-1) in tk^- HeLa and 143 B cells. It was maintained at 20 to 40 copies per cell in HeLa, but at levels of 75 to 120, and rising to several hundred copies per cell in the case of some 143 B cell clones. However, this vector has not been widely used for expressing recombinant proteins.

3.2.4 Adeno-associated virus vectors

Adeno-associated virus-2 (AAV-2) is a human parvovirus which can grow either lytically when co-infected with a helper virus such as adenovirus or HSV, or can persist in an integrated form as a provirus in the absence of helper virus. On superinfection with helper virus the integrated AAV is rescued and a productive infection results. AAV vectors can therefore be used to provide an inducible system in which foreign genes can either be expressed transiently or be integrated into the host genome. The sequences can then be expressed at high copy number by rescue from the proviral state by superinfection (43).

3.2.5 Sinbis virus vectors

Sinbis virus, a single-stranded RNA virus, has been engineered to express the bacterial chloramphenicol acetyltransferase gene (CAT) in cultured insect, avian, and mammalian cells (44). The vectors are self-replicating and express CAT rapidly and efficiently, generating up to 10^8 CAT polypeptides per infected cell in 16 to 20 h. Temperature-sensitive virus mutants are available which may allow the expression of the foreign gene to be regulated by shifting to the non-permissive temperature after infection. Sinbis virus vectors are stable for at least six passages, but are only suitable for short-term use because of the high mutation rate of RNA viruses. At present Sinbis vectors can only be introduced into cells by transfection or micro-injection, but virus-free packaging systems could be used to allow 'suicide' vectors which can infect cells with high efficiency and express the foreign gene without producing infectious virus to be developed.

3.2.6 Retroviral vectors

The properties of recombinant retroviral vectors have been discussed in general terms in Section 2.6. Retroviral vectors in common use fall into two classes: the splicing vectors in which genes are expressed from the endogenous promoter in the retroviral LTR, and the gene of interest and the selectable marker are expressed from differentially spliced transcripts; and the vectors in which the second gene is driven from an internal transcription unit. It has been reported by Emerman and Temin (45) that selection for expression of one gene in this latter type of retroviral vector can result in reduced expression of a second gene in the vector. Thus, selecting infected target cells in G418 would inevitably result in the isolation of clones with reduced levels of expression of the gene of interest. However, results from our laboratory and others indicate that this effect of gene suppression may not be a general phenomenon, and may be limited to vectors derived from avian retroviruses (46).

In spite of the availability of a variety of autonomously replicating viral vectors, there is considerable interest in engineering cells by integrating the heterologous gene into the cellular genome, rather than by expressing it from an episome, because this results in a more stable cell line. High levels of expression can be achieved in two ways: first, by introducing multiple copies of the gene of interest into the cells, for example, by gene amplification; and, second, by regulating expression by choosing strong promoter and enhancer elements which will give high levels of gene expression in the appropriate cell line. Both of these approaches will be considered.

4. Other expression vectors

Many of the mammalian expression vectors in common use are shuttle vectors which can be propagated in both bacterial and mammalian cells. These vectors contain the following basic components: plasmid sequences to allow the vector to be propagated in bacteria; an expression cassette containing a mammalian transcription unit; a selectable marker to enable selection of the cells which have taken up DNA; and sometimes an amplification system for selection for increased copy number.

4.1 Bacterial plasmid sequences

Most mammalian vectors contain sequences from the plasmid pBR322 or its derivative pAT153 including the origin of replication which enables the plasmid to be propagated in bacterial cells. These plasmids contain two selectable markers: the genes for amplicillin and tetracycline resistance, either of which can be used to select for bacteria which have taken up the plasmid. The main difference between the two plasmids is that pBR322 contains sequences (the poison sequences) which have been reported to inhibit replication in certain cell lines, deletion of which in pAT153 results in a marked increase in replication efficiency (47) and hence increased yield of plasmid from a bacterial culture.

4.2 Expression cassette

In addition to the sequence encoding the gene of interest, a number of control elements are needed for mammalian gene expression. At the 5′ end a promoter is required for the initiation of transcription. This consists of a 'TATA box' about 30 nucleotides upstream of the 'cap site' which is where RNA transcription starts. Further upstream, another common sequence element occurs, the CAAT box sequence which is found between 70 and 90 bases upstream of the cap site. Levels of transcription can be increased if enhancer elements are included in the construct. However, these elements often show cell or tissue specificity.

Many promoter sequences have been taken from eukaryotic viruses: these include the retroviral long terminal repeat sequences (LTRs); the SV40 early and late transcription units; the adenovirus major late promoter; the herpes simplex virus thymidine kinase promoter (HSV *tk*); and the human cytomegalovirus immediate early region. Others, such as the mouse metallothionein gene promoter, are derived from cellular genes. Each of these, with the exception of the last promoter, is expressed in a constitutive way, that is, it is active all the time. The metallothionein promoter, in contrast, is an inducible promoter, that is its activity is increased in the presence of a heavy metal such as zinc or cadmium. However, this promoter, like other inducible promoters, e.g. the glucocorticoid promoter, suffers from a high basal level of expression and therefore cannot be regulated as tightly as one would wish.

Although the gene to be expressed is normally included in a vector as a cDNA rather than a genomic sequence, there is evidence to suggest that the presence of introns and their removal by splicing gives rise, at least in some cases, to more stable cytoplasmic transcripts (48). For this reason the SV40 intron and associated splicing signals are frequently included in the expression cassette.

At the 3′ end of the gene there are sequences necessary for termination, cleavage of the 3′ end of the RNA, and polyadenylation, but so far the precise sequences involved have not been well defined. The 3′ end also contains sequences which affect mRNA stability and which must be considered when designing constructs for efficient expression. Kabnick and Housman (49) have shown that, if the 3′ end of a mRNA with a very short half-life (human c-*fos*) is replaced with the 3′ end of the human β-globin mRNA (which has a very long half-life), then the stability of the modified *fos* mRNA is increased. Replacement of sequences in the 5′ untranslated region of globin with part of the 5′ untranslated region of *fos* led to destabilization. It should therefore be possible to improve levels of expression of a heterologous gene by adding mRNA-stabilizing sequences into a construct.

4.3 Selection systems

The uptake of foreign DNA is a very inefficient process and therefore it is important to be able to select those cells which have incorporated and are able to express the foreign DNA. Two types of selection systems exist:

- genes which can only be used in association with mutant cell lines
- genes which act dominantly and can be used in any type of cell

The first category includes genes such as *hgprt, tk*, and *aprt*, which are limited to use with *hgprt*$^-$, *tk*$^-$, and *aprt*$^-$, cell lines, respectively. Although all three types of mutant cell lines are available, *tk*$^-$ and *aprt*$^-$ act as recessive mutations in autosomal genes and therefore both normal copies of the appropriate gene must be inactivated. This requires mutagenesis of the starting cell population and lengthy selection procedures using large numbers of cells. HGPRT$^-$ cell lines can be selected much more easily since *hgprt* is on the X chromosome and therefore only a single copy of the gene is active, but the gene for *hgprt* is large and is less readily available. This type of approach is further limited to systems where it is possible to select for absence of a gene to select for the mutant cell line, and then subsequently for expression of the engineered gene when used as a selectable marker.

This restriction can be overcome by the use of the second category of genes, the dominant-acting genes. One such gene is the bacterial aminoglycoside-3′-phosphotransferase (APH 3′) gene encoded by the bacterial transposon Tn5. Mammalian cells are naturally sensitive to the aminoglycoside antibiotic geneticin (G418), but cells which have taken up and are expressing the bacterial *aph* (or *neo*) gene can phosphorylate, and thus inactivate, the antibiotic. Examples of both types of selectable marker are listed in *Table 3*.

Different mammalian cell lines vary considerably in their inherent sensitivity to G418, and so it is essential to set up a G418 toxicity test before using this selection system. It is important to set this test up at cell densities which are comparable to those which will be used for the selection because the cells are likely to be more sensitive to the drug at plating density than in mass culture.

Table 3. Genes which have been used as markers to select for the expression of foreign DNA

Gene	**Selective agent**	**Reference number**
Require mutant cell		
HSV-1 *tk*	HAT medium	57
aprt	Azaserine/adenine	5, 58
hgprt	HAT medium	58
General applicability		
BPV	Transformed phenotype	39
Ecogpt	Mycophenolic acid aminopterin	59
Tn 601 APH (3′)-1	Geneticin (G418)	60
Tn 5 APH (3′)-11	Geneticin (G418)	61
E. coli hph	Hygromycin B	62
dhfr	Methotrexate	63
asparagine synthetase	β-Aspartylhydroxamate	64
Sh *ble*	Phleomycin	65
cad	PALA	66
Puromycin N-acetyl Transferase	Puromycin	67
E. coli trpB	Indole	68
S. typhimurium hisD	Histidinol	68
bleomycin^{res} gene	Bleomycin	69

4.4 Gene amplification

The level of expression of the foreign DNA in the cell can be increased using by using gene amplification to increase the number of copies of the coding sequence of interest. The system most commonly used is based on the *dhfr* gene which codes for the enzyme dihydrofolate reductase (DHFR) which catalyses the conversion of dihydrofolate to tetrahydrofolate. In 1978 Alt *et al.* (50) showed that one of the mechanisms for mammalian cells to develop resistance to the cytotoxic drug methotrexate was to amplify the gene for *dhfr*. Methotrexate acts by binding to DHFR but, by overproducing the enzyme, a cell can escape from its toxic effect. The cell amplifies the *dhfr* gene and therefore produces excess enzyme so that there is still active enzyme present after all the methotrexate is bound. Selection of cell in progressively increasing concentrations of methotrexate results in the isolation of clones of cells with increasing numbers of copies of the *dhfr* gene.

Using a similar strategy, gene amplification can occur in cells which have taken up a vector containing *dhfr*. In addition to amplifying *dhfr*, other nearby sequences are also amplified: in the case of the methotrexate-resistant mutant cell lines these are the sequences flanking the gene. In the genetically engineered cells the amplified sequences are those adjacent to *dhfr* included in the vector. Thus any gene of interest can be amplified in this way. Best results are obtained if a *dhfr*$^-$ mutant cell line is used as recipient for the gene transfer experiments because amplification of the endogenous cellular gene can be avoided. Using this technique up to 2000 copies of *dhfr* have been reported after three rounds of selection resulting in secretion of in excess of 3×10^8 molecules/cell/day of a co-expressed protein (51).

Other genes which have been used in vectors for co-amplification are listed in *Table 4* and include glutamine synthetase (52) and adenosine deaminase (53). Amplification of the glutamine synthetase (GS) gene occurs when cells are grown in methionine sulphoxamine (MSX), a methionine analogue. This amplification system has the advantage over DHFR in that it can be used equally well in cells which are still expressing the cellular gene.

Table 4. Genes which could be used for selective amplification

Gene	Selective agent	Reference number
Adenosine deaminase	Adenosine/alanosine/deoxycoformycin	53
Multidrug^{res}	Adriamycin	70
P-glycoprotein	Colchicine	71
Ornithine decarboxylase	Difluoromethylornithine	72
Metallothionein 1 & 2	Heavy metals	73
Glutamine synthetase	Methionine sulphoximine	52
dhfr	Methotrexate	63
cad	N (Phosphonoacetyl)-L-aspartate	66
Thymidine kinase (mutant)	HAT	74
hgprt (mutant)	HAT	75
UMP synthase	Pyrazofurin	76
pyr B N(Phosphonoacetyl)-L-aspartate		77
IMP dehydrogenase	Mycophenolic acid	78

4.5 Regulatory sequences

Vectors which rely on amplification to high copy number are often unstable in the absence of selection. This is particularly true of murine cell lines which amplify DHFR sequences as unstable double minute chromosomes which lack a centromere and can be lost from the cell at division. The drugs used for selection, e.g. methotrexate, are frequently expensive and toxic and consequently unsuitable for use in large-scale cultures. There is considerable interest in developing high-level expression vectors in which the elevated level of synthesis is achieved without amplification. Considerable progress in this area is being made, using, for example, the human cytomegalovirus early region promoter which gives high-level expression in a number of mammalian cell lines (54). Friedman *et al.* (55) have reported high-level expression of human growth hormone in CHO cells using the human metallothionein IIA promoter to control expression. Another approach which has potential for the future is to include dominant acting control region (DCR) sequences (56) in the vector in order to overcome reduced levels of expression which occur as a consequence of the site of integration of the vector in the host genome.

5. Conclusions

The development of techniques which allowed genes to be engineered and expressed in novel host cells initially looked likely to reduce interest in making products in animal cells on a commercial basis. If anything, however, the opposite situation has prevailed as the disadvantages of microbial expression systems for the production of human therapeutic proteins have become clearer. Thus, although mammalian cells grow more slowly and to a lower biomass than microbial cells, and although product yields are many times lower than theoretical predictions, nevertheless the comparative ease with which the proteins can be recovered means that genetically engineered animal cells are the source of choice for many products. Current research into the development of improved mammalian expression vectors and a better understanding of mammalian cell growth and product formation in bioreactors seem to suggest an even greater role for engineered animal cells in the future.

Acknowledgements

Part of the work reported in this review was supported by grants from the Science and Engineering Research Council Biotechnology Directorate, the European Community Science Stimulation Programme, and the Humane Research Trust. I should like to thank my colleagues Chris Darnbrough and Pat Watts for their suggestions for improvements to the manuscript.

References

1. Avery, O. T., MacLeod, C. M., and MacCarty, M. (1944). *J. exp. Med.* **79,** 137.
2. Szbalska, E. H. and Szbalska, W. (1962). *Proc. natl Acad. Sci. USA* **48,** 2026.
3. Burnett, J. P. and Harrington, J. A. (1968). *Nature* **220,** 1245.
4. Graham, F. L. and van der Eb, A. J. (1973). *Virology* **52,** 456.
5. Wigler, M., Pellicer, A., Silverstein, S., Axel, R., Urlaub, G., and Chasin, L. (1979). *Proc. natl Acad. Sci. USA* **76,** 1373.
6. Sambrook, J., Fritsch, E. F., and Maniatis, T. (1989). *Molecular Cloning: a Laboratory Manual.* Cold Spring Harbor Laboratory Press, Cold Spring Harbor, New York.
7. Brash, D. E., Reddel, R. R., Quandrud, M., Yang, K., Farrell, M. P., and Harris, C. C. (1987). *Mol. Cell Biol.* **7,** 2031.
8. Wigler, M., Sweet, R., Sim, G.-K., Wold, B., Pellicer, A., Lacy, E., Maniatis, T., Silverstein, S., and Axel, R. (1979). *Cell* **16,** 777.
9. Gorman, C. (1986). In *DNA Cloning: A Practical Approach*, Vol. II (ed. D. M. Glover), p. 143. IRL Press, Oxford.
10. Luthman, H. and Magnusson, G. (1983). *Nucl. Acids Res.* **11,** 1295.
11. Sompayrac, L. M. and Danna, K. J. (1981). *Proc. natl Acad. Sci. USA* **78,** 7575.
12. Stow, N. D. and Wilkie, N. M. (1976). *J. gen. Virol.* **33,** 447.
13. Frost. E. and Williams, J. (1978). *Virology* **91,** 39.
14. Gorman, C. M., Howard, B. H., and Reeves, R. (1983). *Nucl. Acids Res.* **11,** 7631.
15. Felgner, P. L., Gadek, T. R., Holm, M., Roman, R., Chan, H. W., Wenz, M., Northrop, J. P., Ringold, G. M., and Danielsen, M. (1987). *Proc. natl Acad. Sci. USA* **84,** 7413.
16. Scangos, G. and Ruddle, F. H. (1981). *Gene* **14,** 1.
17. Lo, C. W. (1983). *Mol. Cell Biol* **3,** 1803.
18. Yamamoto, F., Furasawa, M., Furasawa, I., and Obinata, M. (1982). *Exp. Cell Res.* **142,** 79.
19. Kurata, S.-H., Tsukakoshi, M., Kasuya, T., and Ikawa, Y. (1986). *Exp. Cell Res.* **162,** 372.
20. Andreason, G. L. and Evans, G. A. (1988). *Biotechniques* **6,** 650.
21. Rassoulzadegan, M., Binetruy, B., and Cuzin, F. (1982). *Nature* **295,** 257.
22. Brown, A. M. C. and Scott, M. R. D. (1987). In *DNA Cloning: A Practical Approach*, Vol. III (ed. D. M. Glover), p. 189. IRL Press, Oxford.
23. Mann, R., Mulligan, R. C., and Baltimore, D. (1983). *Cell* **33,** 153.
24. Miller, A. D., Law, M.-F., and Verma, I. M. (1985). *Mol. Cell Biol.* **5,** 431.
25. Danos, O. and Mulligan, R. C. (1988). *Proc. natl Acad. Sci. USA* **85,** 6460.
26. Rowe, W. P., Pugh, W. E., and Hartley, J. W. (1970). *Virology* **42,** 1136.
27. Eckner, R. J. and Hettrick, K. L. (1977). *J. Virol.,* **24,** 383.
28. Gluzman, Y. (1981). *Cell* **23,** 175.
29. Rigby, P. W. J. (1982). In *Genetic Engineering*, Vol. 3 (ed. R. Williamson), p. 83. Academic Press, London.
30. Muller, W. J., Naujokas, M. A., and Hassell, J. A. (1984). *Mol. Cell Biol.* **4,** 2406.
31. Mackett, M., Smith, G. L., and Moss, B. (1986). In *DNA cloning: A Practical Approach*, Vol. II (ed. D. M. Glover), p. 191. IRL Press, Oxford.

32. Flexner, C., Murphy, B. R., Rooney, J. F., Wohlenberg, C., Yuferov, V., Notkins, A. L., and Moss, B. (1988). *Nature* **335,** 259.
33. Graham, F. L., Smiley, J., Russell, W. C., and Nairn, R. (1977). *J. gen. Virol.* **36,** 59.
34. Berkner, K. L. (1988). *Biotechniques,* **6,** 616.
35. Graham, F. L., Rowe, D. T., McKinnon, R., Bacchletti, S., Ruben, M., and Branton, P. E. (1984). *J. Cell Physiol.,* Suppl. 3, 151.
36. Kwong, A. D. and Frenkel, N. (1985). *Virology* **142,** 421.
37. Desrosiers, R. C., Kamine, J., Bakker, A., Silva, D., Woychik, R. P., Sakai, D. D., and Rottman, F. M. (1985). *Mol. Cell Biol.* **5,** 2796.
38. Cameron, I. R., Possee, R. D., and Bishop, D. H. L. (1989). *Trends Biotechnol.* **7,** 66.
39. Campo, M. S. (1986). In *DNA Cloning: A Practical Approach*, Vol. II (ed. D. M. Glover), p. 213. IRL Press, Oxford.
40. Sugden, B., Marsh, K. and Yates, J. (1985). *Mol. Cell Biol.* **5,** 410.
41. Lupton, S. and Levine, A. J. (1985). *Mol. Cell Biol.* **5,** 2533.
42. Milanesi, G., Barbanti-Brodano, G., Negrini, M., Lee, D., Corallini, A., Caputo, A., Grossini, M. P., and Ricciardi, R. P. (1984). *Mol. Cell Biol.* **4,** 1551.
43. Hermonat, P. L. and Muzyczak, N. (1984). *Proc. natl Acad. Sci. USA* **81,** 6466.
44. Xiong, C., Levis, R., Shen, P., Schlesinger, S., Rice, C. M., and Huang, H. V. (1989). *Science* **243,** 1188.
45. Emerman, M. and Temin, H. M. (1984). *Cell* **39,** 459.
46. Emerman, M. and Temin, H. M. (1986). *Nucl. Acids Res.* **14,** 9381.
47. Luskey, M. and Botchan, M. (1981). *Nature* **293,** 79.
48. Hamer, D. H., Smith, K. D., Soyer, S. H., and Leder, P. (1979). *Cell* **17,** 725.
49. Kabnick, K. S. and Housman, D. E. (1988). *Mol. Cell Biol.* **8,** 3244.
50. Alt, F. W., Kellems, R. E., Bertino, J. R., and Schimke, R. T. (1978). *J. Biol. Chem.* **253,** 1357.
51. Bebbington, C. and Hentschel, C. (1985). *Trends Biotechnol.* **3,** 314.
52. Sanders, P. G. and Wilson, R. H. (1984). *EMBO J.* **3,** 65.
53. Kaufman, R. J., Murtha, P., Ingolia, D. E., Yeung, C.-Y., and Kellems, R. E. (1986). *Proc. natl Acad. Sci. USA* **83,** 3136.
54. Bendig, M. M. (1988). In *Genetic Engineering*, Vol. 7 (ed. P. W. J. Rigby), p. 91. Academic Press, London.
55. Friedman, J. S., Cofer, C. L., Anderson, C. L., Kushner, J. A., Gray, P. P., Chapman, G. E., Stuart, M. C., Lazarus, L., Shine, J., and Kushner, P. J. (1989). *Biotechnology* **7,** 359.
56. Talbot, D., Collis, P., Antoniou, M., Vidal, M., Grosveld, F., and Greaves, D. R. (1989). *Nature* **338,** 352.
57. Colbere-Garapin, F., Garapin, A., and Kourilsky, P. (1982). In *Current Topics in Microbiology and Immunology* (ed. P. Hofschneider and W. Goebel), p. 145. Springer-Verlag, Berlin.
58. Lester, S. C., LeVan, S. K., Steglich, C., and DeMars, R. (1980). *Somat. Cell Genet.* **6,** 241.
59. Mulligan, R. C. and Berg, P. (1981). *Proc. natl Acad. Sci. USA* **78,** 2072.
60. Jimenez, A. and Davies, J. (1980). *Nature* **287,** 869.
61. Colbere-Garapin, F., Horodniceanu, F., Kourilsky, P., and Garapin, A.-C. (1981). *J. mol. Biol.* **150,** 1.

62. Blochlinger, K. and Diggelmann, H. (1984). *Mol. Cell Biol.* **4,** 2929.
63. Wigler, M., Perucho, M., Kurtz, D., Dana, S., Pellicer, A., Axel, R., and Silverstein, S. (1980). *Proc. natl Acad. Sci. USA* **77,** 3567.
64. Andrulis, I. L. and Siminovitch, L. (1981). *Proc. natl Acad. Sci. USA* **78,** 5724.
65. Mulsant, P., Gatignol, A., Dalens, M., and Tiraby, G. (1988). *Somat. Cell mol. Genet.* **14,** 243.
66. de Saint-Vincent, B. R., Delbruck, S., Eckhart, W., Meinkoth, J., Vitto, L., and Wahl, G. (1981). *Cell* **27,** 267.
67. Vara, J. A., Portela, A., Ortin, J., and Jimenez, A. (1986). *Nucl. Acids Res.* **14,** 4617.
68. Hartman, S. C. and Mulligan, R. C. (1988). *Proc. natl Acad. Sci. USA* **85,** 8047.
69. Genilloud, O., Garrido, M. C. and Moreno, F. (1984). *Gene* **32,** 225.
70. Gros, P., Neriah, Y. B., Croop, J. M., and Housman, D. E. (1986). *Nature* **323,** 728.
71. Riordan, J. R., Deuchars, K., Kartner, N., Alon, N., Trent, J., and Ling, V. (1985). *Nature* **316,** 817.
72. McConlogue, L. (1987). In *Gene Transfer Vectors for Mammalian Cells*, p. 79. Current Communications in Molecular Biology. Cold Spring Harbor Laboratory, Cold Spring Harbor, New York.
73. Hamer, D. H. and Walling, M. J. (1982). *J. mol. appl. Genet.* **1,** 273.
74. Roberts, J. M. and Axel, R. (1982). In *Gene Amplification* (ed. R. Schimke), p. 251. Cold Spring Harbor Laboratory Press, Cold Spring Harbor, New York.
75. Kanalas, J. J. and Suttle, D. P. (1984). *J. Biol. Chem.* **259,** 1848.
76. Brennard, J., Chirault, A. C., Karecki, D. S., Melton, D. W., and Caskey, C. T. (1982). *Proc. natl Acad. Sci. USA* **79,** 1950.
77. Ruiz, J. C. and Wahl, G. M. (1986). *Mol. Cell Biol.* **6,** 3050.
78. Collant, F. R. and Huberman, E. (1987). *Mol. Cell Biol.* **7,** 3328.

5

Protein expression and processing

ALAN J. DICKSON

1. Introduction

Understanding the mechanisms which control the generation of proteins from cells in culture is of interest for both commercial and intellectual reasons. For example, the availability of monoclonal antibodies from hybridoma cell lines has great clinical and diagnostic implications. Recent advances in the technology involved in gene manipulation have made it possible to utilize mammalian cells for expression of proteins of defined nature. A number of proteins of possible therapeutic value have been generated from genetically engineered mammalian cells. The procedures for the production of recombinant proteins are designed to ensure the absence of contaminants often found in protein preparations purified from human or mammalian sources by conventional procedures. Several pharmaceutical companies have invested a great deal of time and effort in large-scale production of specific recombinant proteins, including tissue plasminogen activator, blood clotting factors (factors VIII and IX), and erythropoietin (see Chapter 10).

Methods for the introduction and amplification of heterologous genes within host mammalian cells have been dealt with in Chapter 4. The eventual harvest of desired proteins from cells is a reflection of the summation of several processes which lead from transcription of the gene of interest through post-transcriptional processing of RNA, translation, and post-translational processing of the polypeptide chain. In the case of secreted proteins the rate of secretion may be a major determinant towards the capacity to generate the product of interest. Each of the above steps could be a potential limitation towards the generation of protein; as such, each step may require optimization before maximum yield can be obtained. Understanding the mechanisms involved in the control of the above processes in relation to protein formation is crucial for formulation of a protocol to optimize protein yield. For example, for recombinant proteins the inclusion of particular base sequences within the vector used for transfection of the heterologous gene may confer advantages in terms of RNA stability or for the interaction of mRNA with the translational machinery. In addition, the 'fingerprint' of post-translational processing (e.g. proteolysis, glycosylation, acylation) performed in a specific host cell may determine if the recombinant protein is to exhibit full therapeutic value.

The aim of this chapter is to describe the methods available for assessment of protein expression in studies with recombinant cells and other cell lines such as hybridomas. In particular, emphasis will be placed on means for determination of individual stages in the overall sequence of protein formation. I do not seek to provide a comprehensive review of the intricacies of any of the processes involved in the control of protein expression. Instead the aim is to provide a compilation of proven methods which will permit an individual researcher to quantitate the limiting steps involved in control of expression in his or her own system. Information of this type may then provide the starting point for further engineering of cells at the gene level or, in terms of medium constituents, to enhance formation of physiologically active protein.

2. Methods for analysis of protein expression

Assessment of protein expression may require analysis at several different levels. For example, although the synthesis of a specific protein may be identical between two clones of cells arising from a single transfection, there may be differences, and hence limitations, at the level of secretion. Within cells with the same rate of recombinant protein synthesis there could be very distinct contents of mRNA for the specific protein, and synthesis could be limited by different rates of translation. Thus it may prove essential to quantify protein expression at the level of protein amount, at the level of translation, or at the level of transcriptional processes. In the sections which follow I will cover each of these aspects and illustrate the points raised by reference to relevant experimental results.

2.1 Quantitation of protein amount

Within the context of cell culture the presence of protein in most media negates the possibility of a simple protein assay for quantitation of the yield of secreted protein. The release of other proteins by these cells either endogenously or as a consequence of cell lysis would prevent such an approach even in protein-free media. When examining proteins which have to be released by lysis from intracellular sites, these considerations are magnified. Hence, for quantitation of specific proteins one has to turn to the use of either immunological methods or bioassays.

2.1.1 Sample preparation

Even if the protein of interest is destined for secretion, it is advisable to analyse the amount of protein both in media and within cells to assess for the possibility of limitations at the level of secretion. After the desired time period of incubation separation of cells from media is achieved by centrifugation; this should be followed by release of cellular proteins by lysis (either by detergent, sonication, or freezing and thawing cycles). Procedures for this are detailed in *Protocol 1*. It is essential that precautions are taken to minimize loss or degradation of proteins during collection of samples and prior to analysis. These precautions should include the addition of a cocktail of protease inhibitors when collecting proteins from cell samples. This is also advisable for medium samples. All samples should be stored frozen (in most cases at −70°C).

Protocol 1. Separation of cells and medium, cell lysis, and generation of samples for detection of specific proteins

In the following procedures the amounts of lysis buffer used will have to be adapted to the scale of the culture system utilized and the abundance of the material under investigation.

1. For attached cells, remove medium and store. Wash cell sheet rapidly with 25 mM Tris, pH 7.5, containing 20 mM NaCl (wash buffer). Aspirate wash buffer and repeat this wash a further time.

   For cells in suspension separate cells from media by centrifugation (1000 *g* for 5 min). Remove and store medium. Re-suspend cell pellet in wash buffer and repeat centrifugation. Discard supernatant and repeat the wash once more.
2. Lyse the cell samples by addition of wash buffer containing 1% (w/v) sodium deoxycholate, 1% (v/v) triton X-100, 100 μM EDTA, 2 μM E-64, 1 μM leupeptin, 1 μM pepstatin, 200 μM PMSF, and 100 μM TLCK.[a] The protease inhibitors should also be added to medium samples to the same final concentration as given above.
3. Scrape culture dishes with a rubber policeman to ensure complete liberation of all cellular material and transfer this to microcentrifuge tubes. Centrifuge at 10 000 *g* for 5 min and collect supernatant plus any precipitate.
4. Store cellular and medium samples at −20°C (or −70°C) until analysis.
5. As an alternative to detergent lysis it may prove convenient to re-suspend cell pellets in the above extract buffer, lacking the detergent, and to lyse the cells by three cycles of freezing (liquid nitrogen) and thawing (37°C).

[a] Other detergents may be used, e.g. 0.5% Nonidet NP40.

2.1.2 Analytical procedures

Materials generated from cell or medium fractions of cell culture as outlined in *Protocol 1* may be used for analysis of specific protein content. Assay methods can be divided into two types:

- functional or bioassay types
- immunological

It is difficult to generalize on the method for a bioassay; although molecules may have a defined biological function which can be assessed in relation to standards, the ease with which bioassays can be formulated is variable and not all proteins exhibit a measurable function. For examples of such techniques the reader is advised to examine the cases of interleukins and interferons in an earlier volume in this series (1). In terms of sensitivity there is little difference between bioassays and immunological assays. In the case of gamma interferon, for example, both types of assay detect to 2–20 units interferon/ml (1). Immunologically based assays are thought of as sensitive and easier to perform with multiple samples. It should be pointed out that, in the case of recombinant proteins which are being synthesized in an engineered cell, protein may be detected by immunological methods but may not be detected by a bioassay. For example, specific antibodies may still recognize proteins which have been partially- or mis-glycosylated and which, as a consequence, may fail to exhibit biological function. Some immunological detection systems are based on radioimmunoassay methods (1). However, there are non-radiometric systems which can be easier to perform. The first involves detection based on enzyme-linked immunoabsorbant assays (ELISA) which can be used in a quantitative analysis of protein amount. A protocol for a typical specific assay which measures the functional ability of murine immunoglobulin to recognize an antigen is detailed in Chapter 6. Non-specific assays also exist which involve the binding of immunoglobulin in culture medium to microtitre plates and subsequent analysis of immunoglobulin amount by interaction with antibodies which recognize the constant regions of immunoglobulin heavy chains. The second method involves detection by immunoblotting which is a more qualitative indicator of the degree of protein production.

Immunoblotting is a non-radiometric method which offers advantages for rapid screening of clones of cells for comparisons (*Protocol 2*). The technique allows rapid assessment of differences in protein expression (*Figure 1*) and also of possible post-translation modifications to proteins. Using antibodies against the phosphorylated epitope only or against both phosphorylated and dephosphorylated forms of phenylalanine hydroxylase, Green *et al.* (2) have shown that immunoblotting can define the state of phosphorylation of a specific protein. This approach has immense promise for screening for post-translational modifications of proteins.

Protocol 2. Immunoblotting analysis of protein contents

Reagents and apparatus

- Electrotransfer apparatus: either commercial (e.g. LKB Multiphor II) or home-built

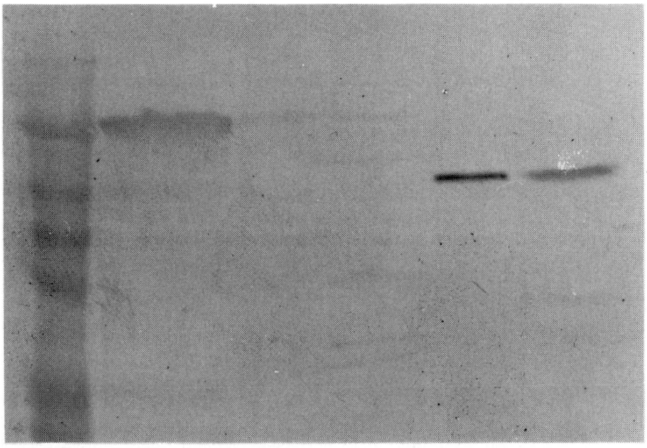

Figure 1. Immunoblot analysis of tissue content of protein kinase-C. Extracts were prepared from fed adult rat brain or liver, as indicated. Homogenates were prepared (*Protocol 1*) and used directly or separated into cytosolic and membrane fractions. Portions (60 μg) of each fraction were applied to, and separated on, SDS–PAGE (see *Protocol 6*). After transfer to nitrocellulose, filters were blotted for the presence of the β_2-isoenzyme of protein kinase-C (M_r 78 000–82 000) with an isoenzyme-specific antiserum provided by Dr M. Merrett, Wellcome Research Laboratories, Beckenham, Kent, UK. Lane 1: Prestained molecular standards, subunit molecular weight in brackets; α_2-macroglobulin (180 000), β-galactosidase (116 000), fructose-6-phosphate kinase (84 000), pyruvate kinase (58 000), fumarase (48 500), lactate dehydrogenase (36 500), triose phosphate isomerase (26 000); lane 2: unfractionated liver extract; lane 3: liver cytosolic fraction; lane 4: liver membrane fraction; lane 5: brain cytosolic fraction; lane 6: brain membrane fraction.

- Nitrocellulose: from various suppliers, e.g. Gelman
- Blotting buffer: 3 g Tris base and 14.4 g glycine added to 780 ml water, pH is adjusted to 7.4 and volume to 800 ml. Add 200 ml methanol to the mixture
- TBS/Tween: mix 1.211 g Tris base, 8.18 g NaCl and 1 ml of Tween 20. Adjust final pH to 7.4 and volume to 1 litre
- Blocking buffer: add 5 g casein milk powder (we use Marvel) to 100 ml TBS/Tween; re-adjust pH to 7.4
- BCIP: make up 5-bromo-4-chloro-3-indolylphosphate *p*-toluidine in dimethyl formamide (to 50 mg/ml). Store at −20°C
- NBT: make up nitroblue tetrazolium chloride in 70% dimethyl formamide (to 50 mg/ml). Store at −20°C
- Development buffer: mix 1.211 g Tris base, 0.584 g NaCl and 0.101 g $MgCl_2{\cdot}(H_2O)_6$ and adjust pH to 9.5 and volume to 100 ml. For 10 ml of buffer add 33 μl BCIP and 66 μl NBT immediately prior to use

Protocol 2. *continued*

- Primary antibody: this will have been raised against the antigen to be detected. For a good reaction it should recognize both native and denatured antigen. The optimal working dilution must be determined in preliminary experiments but will lie between 1:200 and 1:1000.
- Secondary antibody: this should be directed against the primary antibody, i.e. if the primary antibody was from rabbit use anti-rabbit immunoglobulin. Dilute to 1:1000 for working stocks. We use anti-rabbit immunoglobulin-alkaline phosphatase conjugate.[a]

Methods

1. Run the samples containing the protein of interest on slab SDS–PAGE (we run 0.75 mm thick gels). Remove the gel and soak in blotting buffer for 30 min until the Coomassie blue dye diffuses from the gel.
2. Sandwich together the gel and a piece of nitrocellulose of the same size between three layers of filter paper on each side. This should be done in blotting buffer to ensure that there is thorough wetting of the nitrocellulose. Roll out any air bubbles from the 'sandwich' and set up in the electrotransfer apparatus. Precise time of transfer depends on model and gel size. We routinely run pre-stained SDS–PAGE molecular weight standards on every gel to assess the completeness of transfer.
3. At this stage the nitrocellulose can be stored overnight in TBS/Tween at 4°C. The subsequent steps are performed in 10 cm^2 plastic dishes.
4. Cover the nitrocellulose with blocking buffer and place on an orbital shaker for 30 min. Remove buffer and replace with fresh blocking buffer containing the primary antibody at the optimal dilution. Shake at room temperature for 30–60 min.
5. Remove the buffer and wash the nitrocellulose with fresh blocking buffer. Repeat washing four more times, leaving each wash for 5–10 min with shaking.
6. Add the secondary antibody diluted 1:1000 in blocking buffer and incubate for 30–60 min with shaking. Wash as before three times and then twice with TBS/Tween.
7. Add development buffer and shake gently until colour develops. Good quality reactions will appear rapidly, i.e. within 30–60 sec.

[a] Other secondary antibodies can be used, e.g. horseradish peroxidase with consequent effects on substrates for the development buffer.

Other immunological methods exist for detection of proteins. For example, following immunoprecipitation in tubes, the products can be run on SDS–PAGE. After staining for proteins gels can be scanned and estimates obtained for protein amounts (see *Protocol 6*).

2.2 Quantitation of translational and post-translational events

Prior to the appearance of the final protein product several steps must occur:

- formation of mRNA/ribosomal complexes
- initiation and maintenance of translation
- co-translational and post-translational processing
- movement of newly synthesized protein to the correct cell compartment for further processing or secretion

Assessment of the individual steps in the complex interactions that comprise translational and post-translational events generally involves the use of radioisotopes for monitoring the metabolic fates and rates of interconversion of proteins. Modifications of proteins can be essential for targeting to particular sites in the cell relevant to the rate of secretion and also for full expression of their physiological function in terms of clearance rate from serum or targeting to specific tissues (3, 4). Examples include the differences between the same polypeptide subject to varied degrees of glycosylation when function is assessed by bioassay (5, 6). In addition, there is the possibility of microheterogeneity within the glycosylation pattern of polypeptides. This has been observed for gamma interferon (7); however, although the physiological validity of this is unknown, the pattern of glycosylation may be important for assessment of the quality of a given recombinant protein.

Other modifications which may occur co- or post-translationally include proteolysis of a precursor polypeptide (7), formation of disulphide bonds, and the addition of acyl or sulphate groups. Proteolysis or specific disulphide bond linkage may maintain the 'correct' protein conformation for function or localization. Although sulphation of cellular proteins is common, it is unknown what function this may offer for the cell or protein although charge may again determine function or localization of the protein. Acylation to specific amino acids is essential for the localization of several oncoproteins to the plasma membrane (8). Knowledge of the extent of such modifications for any given polypeptide will be important as several can determine the partitioning of proteins between distinct cellular compartments and thus may control secretion or retainment. Although radioisotopic methods are generally favoured for studying the processing events given above, it should be noted that use of antibodies which selectively recognize epitopes produced after processing may be more convenient for assessment of individual processing steps (2).

In the following section I will describe two systems to examine translational and post-translational events:

- intact cell systems
- cell-free or reconstituted systems

There are advantages to working with the simplified cell-free system although these systems are difficult to optimize. For this reason I will deal with these briefly at the end of this section.

2.2.1 Translational processes in intact cells

In these procedures cells are incubated in standard media to which is added radioactively labelled amino acid to assess translation (*Protocol 3*). It is most common to use ^{35}S-methionine or ^{3}H-leucine as the amino acid label (*Figure 2*); however, there may be reasons for choosing other labelling regimes when a specific amino acid is prevalent in the desired protein product, e.g. the use of ^{3}H-proline in the measurement of synthesis of collagens (9).

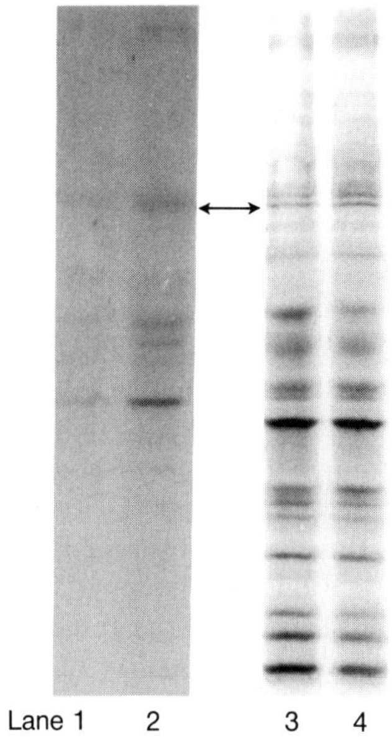

Figure 2. Synthesis of total proteins and phosphofructokinase-1 in response to epidermal growth hormone (EGF). Chick embryo hepatocytes were cultured in serum-free Waymouths MB 752/1 for 24 h in the presence of [^{35}S]methionine (final specific activity, 20 μCi μmol). Cells were incubated throughout this period in medium alone (lanes 1 and 3) or in the presence of 10^{-8} M EGF (lanes 2 and 4). At the end of the culture period, medium was removed from the cells and the cells were extracted (*Protocol 1*) to produce a cytosolic extract. Samples of extracts containing equivalent amounts of radioactivity were subjected to SDS–PAGE directly (lanes 3 and 4) or incubated with an antibody raised against chick liver phosphofructokinase-1. After reaction with the antibody, the immunoprecipitate was analysed on a separate SDS–PAGE gel (lanes 1 and 2). Gels were subsequently vacuum-dried and used to generate the autoradiographs which are presented here. The arrow indicates the position where phosphofructokinase-1 (87 000–89 000) is found relative to molecular weight standards run simultaneously.

Protocol 3. Measurement of translational capacity of intact cells

Materials

- Radiolabelled amino acid, e.g. L-[^{35}S]methionine, L-(4,5-^{3}H]leucine)
- 10% (w/v) trichloroacetic acid (TCA)
- 5% (w/v) trichloroacetic acid containing 10 mM unlabelled amino acid, e.g. methionine or leucine as relevant to radiolabel
- NCS tissue solubilizer (Amersham, UK) or similar solubilizer

Methods

1. For each time point of radiolabel incorporation a sample corresponding to 5–10 × 10^5 cells should be analysed. For cells in suspension they should be incubated at 5–10 × 10^6 cells/ml. For attached cells each flask or test sample should contain a total of 5–10 × 10^5 cells.
2. Initiate the experiment by addition of radiolabelled amino acids at final specific activities of 20–40 μCi/ml. Continue incubations in conditions which are normal for the cells under study. Perform incubations over periods up to 4–6 h.
3. Terminate reactions at desired time intervals as given below with separation of cellular and medium samples being performed prior to termination. Take a sample at zero time to assess the background in the absence of incorporation.
4. Separation of cellular and medium samples:
   (a) For attached cells remove all the medium and centrifuge the medium at 10 000 *g* for 1 min. Wash the cell sheet with wash buffer (see *Protocol 1*). Discard the wash. Deal with the medium supernatant and cellular fractions as detailed below[a].
   (b) For cells in suspension remove a portion (50–100 μl) of the homogeneous suspension into a 1.5 ml microcentrifuge tube and centrifuge (10 000 *g* for 1 min) to separate cells and medium. Remove the medium fraction and store. Wash the cell pellet gently with 500 μl of wash buffer (*Protocol 1*), repeat the centrifugation, and discard the supernatant. Deal with cellular and medium fractions as below.[a]
5. Add medium samples (50–100 μl) to an equal volume of ice-cold 10% TCA in a 1.5 ml microcentrifuge tube. Complete protein precipitation occurs when the tubes are left on ice for 10–15 min. For cellular samples precipitate protein by addition of 500 μl of 5% TCA containing excess unlabelled amino acid (methionine or leucine) to the pellet of cells in the microcentrifuge tube; with attached cells make the addition to the culture dish and scrape the material off into a microcentrifuge tube. Subsequent processing is the same for medium and cellular fractions.[b]
6. Re-suspend the protein precipitate in 500 μl 5% TCA containing unlabelled

Protocol 3. *continued*

amino acid and heat at 90°C for 15 min to hydrolyse any charged tRNA. Centrifuge at 10 000 *g* for 2 min.

7. Wash the protein precipitates three times by re-suspension in 500 μl 5% TCA containing the respective unlabelled amino acid and re-centrifuge (10 000 *g*, 2 min). Remove the liquid from the precipitate and process by one of two methods.
   (a) Measurement of total incorporation. Add 30 μl of NCS tissue solubilizer to the pellet. Stir and leave to dissolve for 60 min at room temperature. Cut entire tip from tube and count the tip and its contents in the presence of 2 ml of Ecoscint A (or similar scintillation fluid) plus 10 μl 10 M HCl.
   (b) Measurement of incorporation into specific proteins. Add SDS–PAGE sample buffer and separate into polypeptide subunits by SDS–PAGE followed by autoradiography, fluorography, or slicing of gel into segments to identify incorporation into specific proteins (see *Protocol 6* for details).

[a] As an alternative, material can be processed without acid precipitation for direct comparison of amino acid incorporation into specific proteins and total proteins (see *Protocol 4*).

[b] As an alternative to the above, aliquots of medium proteins may be portioned on to Whatman no. 1 paper squares (1 cm^2). The papers are placed in 10% TCA and go through washing procedures similar to the above before being dried and subjected to scintillation counting.

The fate of incorporated radiolabel can identify changes to specific proteins either by separation of protein products on SDS–PAGE in isolation (*Protocol 6*) or after isoelectric focusing. Although some information can be gathered from the general pattern of labelled polypeptide species from electrophoretic analysis, much more defined data comes from use of immunoprecipitation (*Figure 2*) to detect the incorporation of amino acid into specific proteins (*Protocol 4*).

Protocol 4. Identification of specific translation products by immunoprecipitation

Materials

- Good quality antibody specific to the antigen of interest: this can be monoclonal or polyclonal[a] but the titre must be characterized.
- Wash buffer and wash buffer containing protease inhibitors (see *Protocol 1*)
- Wash buffer and protease inhibitors at twice normal working concentrations

Methods

1. Incubate cell cultures with radiolabelled amino acids as for *Protocol 3* (up to stage **4**). After separation of cells from medium, lyse cells in wash buffer containing protease inhibitors as given in *Protocol 1*. Mix medium samples with an equal volume of double-strength wash buffer containing protease inhibitors. At this stage samples can be stored frozen.
2. Mix portions of the detergent-treated cell or medium samples with sufficient antibody to immunoprecipitate two to three times the amount of antigen suspected as being present in the samples. Incubate for 1–2 h at 37°C with gentle shaking and subsequently at 4°C overnight with constant rotation if possible.
3. Collect immunoprecipitates by centrifugation (10 000 *g*, 4°C, 5 min)[a].
4. Wash the pellets by three cycles of re-suspension in wash buffer containing protease inhibitors plus 0.1% (w/v) SDS followed by re-centrifugation (as stage **3**).
5. *Either* re-suspend pellet in 30 μl NCS (see *Protocol 3*) and count directly for incorporation of radioactivity *or* re-suspend in sample buffer prior to analysis of composition by SDS–PAGE (see *Protocol 6*).
6. Assessment of non-specific precipitation can be made by:
   (a) repeating the analysis by addition of a further portion of antiserum to extracts which have already undergone complete immunoprecipitation.
   (b) using pre-immune serum on a duplicate portion of extracts.

[a] For immunoprecipitation of small amounts of antigen with polyclonal antibodies it is possible to enhance complex formation, and hence precipitation, by addition of partially-purified antigen (carrier antigen) to the extracts. For most classes of monoclonal antibodies the addition of protein A–sepharose or protein A–agarose following the initial incubation with antibody (stage **2**) is required to effect precipitation. This technique is also effective with protein G and addition of either will speed up complete precipitation with polyclonal antibodies.

Several commercial companies can provide media lacking the single amino acid selected for use as radioisotopic label. This type of medium is used for 'pulse-chase' experiments which permit the identification of the sequence and timing of processing events which lead to the final protein product (*Protocol 5*). Here, cells are cultured in medium with high specific activity of the radioactive amino acid (i.e. in medium where the radioisotope is the only form of that amino acid) for a short period of time (minutes rather than hours). Removal of that medium and replacement with medium which contains high concentrations of that particular amino acid in a non-radioactive form permits the estimation of the secretion rate and the identification of radiolabelled intermediates in the sequence of cellular intracellular modifications of proteins (5).

Protocol 5. Outline of pulse-chase procedure

This protocol should be read in parallel with *Protocol 4* as there are several common steps.

1. Culture the cells initially in the absence of added radiolabelled amino acids.[a]
2. Initiate the pulse-chase experiment by changing the medium to a formulation which is normal for the cells under investigation with the exception of the lack of the amino acid to be used as label, e.g. methionine or leucine. This medium should be supplemented with labelled amino acid, e.g. L-(^{35}S]methionine or L-[4,5-^{3}H]leucine to give a final content of 50–100 μCi/ml (pulse-medium).
3. Incubate the cells in this medium for a short time interval, sufficient to allow equilibration of label with newly synthesized proteins, i.e. 5–60 min.
4. Remove pulse medium and wash cells gently with complete medium lacking radiolabelled amino acid but containing that amino acid in an unlabelled form at 2 mM (chase medium). The wash is performed twice.
5. Add fresh chase medium to each flask and incubate under normal conditions of culture for an interval of 4–6 h. Samples are assessed for changes in labelling at frequent time intervals (15–30 min), which should be characterized in preliminary experiments. At each time interval examine
   - medium samples
   - cellular samples

   Preparation of cellular and medium samples and assessment of radioactivity in total or specific proteins are outlined in *Protocols 1, 3*, and *4*. Although sequential time samples can be removed from the medium of both suspension and attached cell lines, for analysis of cellular material with attached cells it is necessary to use a single flask for each time point.
6. A zero 'reference' point is obtained by analysis of material at the start of the chase period.
7. Measurement of TCA-soluble material (see *Protocol 3*) in the medium during the chase period can be used as an estimate of total protein degradation under the conditions of culture (10).

[a] Volumes of medium used in these experiments should be related to the dimensions of the culture dishes or flasks used but generally the volume should be kept to a minimum.

In *Protocol 5* it is possible to monitor the changes which occur to specific proteins through use of antisera in a manner analogous to that described in *Protocol 4*. However, it is also possible to examine changes in overall translational capacity or in the processing of total cellular polypeptides by use of

SDS–PAGE to separate polypeptides and then by analysis of radioactivity within distinct polypeptide species (*Protocol 6*).

Protocol 6. Separation and identification of translation products by SDS–PAGE

Materials

- Separating buffer: weigh 45.4 g Tris base (Sigma) and 1 g SDS (sodium lauryl sulphate, electrophoresis grade). Adjust pH to 8.8 and final volume to 250 ml. Store at room temperature
- Stacking buffer: weigh 15 g Tris base and 1 g SDS. Adjust pH to 6.8 and final volume to 250 ml. Store at room temperature
- Sample buffer: weight 0.76 g Tris base, 2 g SDS, and 5 mg bromophenol blue and mix with 35 ml of water and 10 ml of glycerol. Adjust pH to 6.8 and final volume to 50 ml. Store at room temperature and add 18 μl 2-mercaptoethanol/ml just prior to use
- Electrode buffer: weigh 12.02 g Tris base, 4 g SDS, and 57.68 g glycine and make up to a final volume of 2 litre. Store at room temperature.
- Acrylamide stock: weigh 30 g acrylamide and 0.8 g N,N′methylene bis-acrylamide and make up to a final volume of 100 ml. Filter through a 0.45 μm filter and store in the dark at 4°C. Use within 1 month of preparation.
- Ammonium persulphate: make up 100 mg of ammonium persulphate in 1 ml water just before use.
- TEMED: stock bottle of N,N,N′,N′-tetramethylethylenediamine required.

Methods

1. Use a gel system based on a separating gel overlaid by a stacking gel containing 4% acrylamide. The concentration of acrylamide in the separating gel will determine the effective separation range for polypeptides. A separating gel with 10% acrylamide will be effective over a molecular weight range of 16 000 to 75 000, whereas a 5% separating gel will be effective from 60 000 to 200 000.
2. The details which follow are for the Biorad Mini Protean 2 slab gel system; however, the amounts can be scaled to suit any other slab or stick gel system.
3. Prepare 10 ml of a 10% acrylamide separating gel mix by mixing together:
    - 3.3 ml acrylamide stock
    - 2.5 ml separating buffer
    - 4.2 ml water

Protocol 6. *continued*

Initiate polymerization by addition of 100 μl of ammonium persulphate and 10 μl TEMED.

4. Immediately pour separating gel mixture into gel slab, leaving space at the top for the stacking gel (1 cm) and for the Teflon comb to make the sample wells. Overlay the separating gel gently with isobutyl alcohol. When the gel has set (15–20 min), pour off the alcohol and rinse the top of the gel with water.
5. Prepare 10 ml of a 4% acrylamide stacking gel mix by mixing together:
   - 1.32 ml acrylamide stock
   - 2.5 ml stacking buffer
   - 6.18 ml water

   Initiate polymerization by addition of 100 μl of ammonium persulphate and 10 μl TEMED.
6. Immediately pour stacking gel mix into the gel sandwich on top of the separating gel. Before this gel sets place the Teflon comb into stacking gel and allow to set (15–20 min).
7. Prepare samples or standards by mixing equal volumes of protein solution with sample buffer. Heat the mixtures at 100°C for 5 min and allow to cool before placing on the gel. Up to 100 μg of an impure protein preparation or 30 μg of a pure protein in a total volume of 15–20 μl can be loaded in each well. Standard proteins as molecular weight markers for SDS–PAGE are available from a number of commercial suppliers.
8. When the stacking gel has set remove the comb and place the gel sandwich in the gel tank. Cover cathode and anode reservoirs of the tank with electrode buffer and apply samples or protein molecular weight standards to wells in the stacking gel.
9. Run the electrophoresis at 60 V until the bromophenol blue marker reaches the interface between stacking and separating gels then increase and maintain at 200 V until the marker dye is near the end of the separating gel.
10. At the end of the run remove the gel from the plates and carry out one of the following:
    (a) Stain the gel for protein bands. Immerse the gel in a mixture of methanol: acetic acid: water (4.5:1:4.5) (containing 1 mg Coomassie blue R250 ml), normally overnight but this can be less. Destain in the same solution minus Coomassie blue until bands are visible with little background.
    (b) Vacuum-dry the gel and perform autoradiography or carry out fluorography.

(c) After staining cut the gel lanes into 1–2 mm slices and solubilize these by incubation in 200 μl of 20 vol. H_2O_2 at 70°C for 5–18 h in scintillation vial insets. Subsequently, count the radioactivity in each slice.

For examination of specific proteins there may be other methods to isolate the protein of interest without resorting to immunological methods prior to its analysis by direct counting or after electrophoresis. Examples include the purification of glycoproteins by lectin-based affinity columns (5) and the use of immobilized protein A or G in the determination of secretion of several classes of immunoglobulin by hybridoma cells (*Protocol 7*).

Protocol 7. Determination of immunoglobulin secretion in murine hybridoma cells: purification with protein A

This technique is applicable for detection of the rates of immunoglobulin synthesis and secretion using the radiolabelling procedures detailed in *Protocol 3*. The outline below describes analysis of medium; however, this could be applied to intracellular material (see *Protocols 1* and *3*).

Materials

- Wash buffer (see *Protocol 1*)
- Anti-mouse immunoglobulin antiserum (IgG type)
- Protein A–sepharose. Prepare a solution by re-suspending commercial protein A–sepharose in wash buffer, centrifuge at 5000 *g* for 20 sec, and re-suspend in wash buffer at the desired concentration.
- 50 mM sodium citrate, pH 5.5
- 50 mM sodium acetate, pH 4.3

Methods

1. Aliquot media (100 μl) from hybridoma cell cultures into microcentrifuge tubes and add 50 μl rabbit anti-mouse immunoglobulin antiserum (Sigma, 0.115 mg/50μl).
2. Incubate at 37°C for 1–2 h then add 20 μl (20 μg) protein A–sepharose complex. Incubate at room temperature with constant shaking (wheel mixer) for 40 min.
3. Collect complex by centrifugation (10 000 *g*, 1–2 min). Remove supernatant and successively wash the pellet with 250 μl wash buffer, 250 μl citrate buffer, and 250 μl acetate buffer.

Protocol 7. *continued*

4. Analyse the pellet in one of the following ways.
   (a) Solubilize the precipitate and directly count incorporation of radioactivity.
   (b) Solubilize the materials for examination on SDS–PAGE by protein staining or radioactivity detection (see *Protocol 6*).

2.2.2 Post-translational processes in intact cells

Protocols 3 to *7* can be modified to produce methods to examine post-translational modifications in intact cells. Data on proteolysis can be gained by 'pulse-chase' experiments coupled to the use of immunoprecipitation and analysis of product size by SDS–PAGE. To assess the extent of other modifications, the radioactive amino acid added to culture media can be replaced with:

- radioactive sugars (for measurement of glycosylation)
- radioactive fatty acids (for measurement of acylation)
- radioactive inorganic sulphate (for measurement of sulphation)

The modifications which occur to any specific protein can be assessed by the methods outlined in *Protocols 4* and *5* and pulse-chase experiments can be adapted to radiolabelled compounds other than amino acids. At a gross level it is also possible to detect differences in glycosylation of specific proteins by two simple methods.

- lectin affinity chromatography followed by identification of binding affinity of immunoreactive species (5)
- lectin blotting following SDS–PAGE of total or affinity- or immuno-purified proteins (11)

It is beyond the scope of this review to suggest methods for analysis of structures within glycosylation sites or of localization of sequences within proteins which are subject to other specific modifications. Two recent papers serve to illustrate the current status of technology available for analysis of carbohydrate structures at glycosylation sites (12, 13).

2.2.3 Cell-free systems

The situation within the intact cell involves complex interactions requiring several cell compartments. To identify the factors which are relevant the control of translation or translocation into the endoplasmic reticulum or to post-translational modifications, several research groups have attempted to use cell-free systems which allow easy manipulation of conditions. Although cell-free translation from the rabbit reticulocyte lysate has long been a standard procedure, there has been great difficulty in production of active cell-free translation systems from other cell types. Data from liver, for example, suggests that these systems are a long way short of physiological relevance (14). Other workers have attempted to set up protocols to examine controlling factors in systems which give coupled synthesis, secretion, and processing. There has been some success on this front (7) but there is no universal system available. If, or when, such systems become applicable to cells in general, *Protocols 3* to *7* can be modified to be applicable.

2.3 Quantitation of transcriptional processes

Description of the complex events involved in initiation and maintenance of transcription and in post-transcriptional processes is beyond the scope of this review. In addition, it is unlikely that with the current state of knowledge it would be possible to give a full account of all relevant steps. At present it is possible to assess the rates of transcription from the viewpoint of mRNA amount and the rates of formation and degradation of mRNA. In order to examine these parameters as they relate to a specific gene product it is crucial to have a specific nucleic acid probe against the product of interest. When considering recombinant proteins this is unlikely to be a problem as the isolation and use of the recombinant gene will, in addition, allow the isolation of cDNA or genomic probes for the gene whose expression is under examination.

In the sections which follow I will discuss how detection of specific

- mRNA content
- rates of gene transcriptional activity

can be used to focus on to the aspects of protein expression which may limit final protein production.

2.3.1 Isolation, detection and quantitation of mRNA

Several methods have been utilized for isolation of RNA. I have found that the CsCl/guanidinium method based on that of Chirgwin *et al.* (15) gives reproducible high-quality results even for many tissues such as liver which express high levels of ribonuclease (*Protocol 8*). Other methods may be more appropriate for certain other tissues; these are reviewed elsewhere (16, 17).

Protocol 8. Isolation of RNA for analysis of specific mRNA contents

Precautions must be taken to ensure that all solutions which come into contact with RNA are free of ribonuclease activity; the use of diethyl pyrocarbonate (DEPC) to inactive ribonuclease is essential (16, 17). All solutions below are prepared with DEPC-treated water[a] and are autoclaved before use.

Materials

- 50 mM sodium citrate: 0.735 g made to a final pH of 7.0 in 50 ml.

Protocol 8. *continued*

- 20% (w/v) sarkosyl: 2 g sarkosyl dissolved in 10 ml
- Extraction buffer: 10 ml sodium citrate, 2.5 ml 20% sarkosyl, 47.28 g guanidinium isothiocyanate, 0.1 ml antifoam A, and 0.7 ml 2-mercaptoethanol made up to 100 ml
- 0.1 M disodium EDTA: dissolve 3.722 g EDTA and adjust pH to 7.5 and final volume to 100 ml
- 20% SDS: dissolve 20 g SDS in 100 ml
- 3 M sodium acetate: dissolve 40.824 g of the trihydrate salt and adjust pH to 5.2 and volume to 100 ml
- 5.7 M caesium chloride: 47.985 g caesium chloride dissolved in 50 ml 0.1 M EDTA
- 10 mM Tris containing 5 mM EDTA and 1% SDS: mix 60.55 mg Tris base, 93.06 mg EDTA, and 2.5 ml 20% SDS and make up to a final volume of 50 ml

Methods

1. Perform cell culture in the normal manner and at the required time remove the medium from the cell sheet (for attached cells) or separate cells from medium by centrifugation (suspension cells, see *Protocol 1*). 5×10^7–10^8 cells should be utilized for preparation of RNA.
2. Add 9 ml extraction buffer to each cell sample. Leave for 10–20 min to allow complete lysis of cells. Transfer extracts to sterile universal tubes. Add 4 g solid caesium chloride to each extract.
3. Place a 2.5 ml cushion of 5.7 M caesium chloride in a heat-sealable centrifuge tube (14 ml total volume). Apply cell extract on to cushion through a 10 ml syringe and needle. Seal the tubes and centrifuge in a fixed angle rotor at 52 000 *g* for 16 h at 18°C.
4. Remove the top layer of liquid with a syringe, cut the top off the tube, and decant off any remaining liquid without disturbing the pellet. Remove excess liquid from walls of the tube.
5. Dissolve the pellet in 0.5 ml 10 mM Tris containing 5 mM EDTA and 1% SDS; ensure this is completely dissolved (30 min). Transfer liquid to sterile microcentrifuge tubes and extract the sample with 0.5 ml phenol: chloroform: isoamylalcohol (25:24:1).
6. Centrifuge at 10 000 *g* for 5 min and remove the aqueous (top) layer into a fresh microcentrifuge tube. Re-extract with 0.5 ml chloroform:isoamylalcohol (24:1). Repeat centrifugation and remove top phase into a fresh microfuge tube.
7. Add 50 μl 3 M sodium acetate to the collected top layer and fill the tube to the top with absolute ethanol. Mix and place at −20°C for at least 2 h.

Centrifuge at 10 000 *g* for 10 min at 4 °C and remove ethanol. Rinse the pellet with 70% ethanol; then freeze-dry the pellet to remove all traces of ethanol.

8. Dissolve the pellet in the least possible amount of DEPC-treated water. This takes at least 30 min. RNA can now be analysed or stored at −70 °C.

[a] DEPC-treated water is prepared by adding 0.1% (v/v) DEPC to distilled, deionized water. The treated water is left overnight and autoclaved the next day.

The yield of RNA is assessed by measured of A_{260}; one absorbance unit at this wavelength is equivalent to a concentration of 40 μg/ml. In addition, measurement of the ratio of A_{260}:A_{280} gives an estimate of the contamination of the RNA preparation by protein. Values of 1.75 to 1.85 are routinely obtained with this procedure. Typically, this procedure yields 200–300 μg total RNA from 10^8 cells.

When determining the amount of a specific mRNA within the total population of RNA with a nucleic acid probe, I would not routinely purify mRNA on affinity columns such as oligo-dT cellulose or poly-U sephadex. Such column chromatographic procedures add a further step in which RNA degradation can occur and, for most relatively abundant mRNA species, the sensitivity of nucleic acid hybridization techniques requires no concentration of mRNA. If necessary, the specific activity of the nucleic acid probe can be increased by different labelling procedures. In increasing order of sensitivity (16, 17) these are:

- nick translation
- random priming
- riboprobing

In addition, it is possible that the affinity columns select only a subpopulation of mRNA (i.e. that with relatively large tracts of poly A tails) and this might not be a reflection of the total transcriptional activity of cells.

For detection of specific mRNA species I would always recommend the use of Northern blotting (in which different mRNA species are separated by agarose gel electrophoresis of total RNA) rather than the techniques of dot- or slot-blotting. Following electrophoresis, RNA is transferred from the gel on to nitrocellulose or nylon filters; this is then probed for the mRNA of interest by hybridization using a nucleic acid probe. Detection is based on the presence of an identifiable label on the probe, e.g. ^{32}P or luminescent material (16, 17). With prior electrophoresis of RNA samples there are two distinct advantages.

(a) With ethidium bromide staining the appearance of the ribosomal RNA species is a rapid indicator of RNA integrity (*Figure 3*).

(b) The possibility of multiple mRNA types for the species of interest which arise from differential processing can be determined.

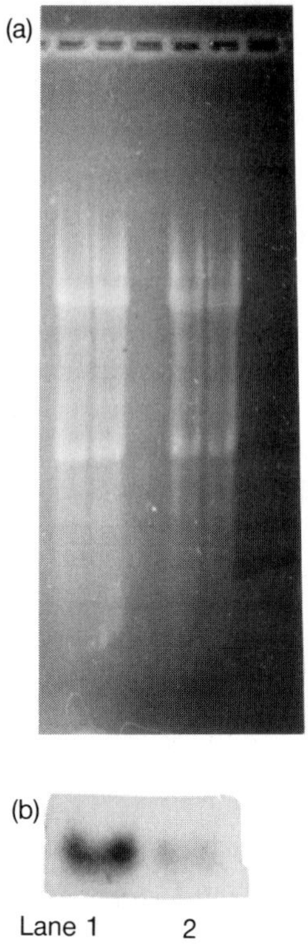

Figure 3. Isolation of intact RNA for detection of specific mRNA species by Northern blotting. RNA was isolated from cultured rat hepatocytes and separated on agarose gel electrophoresis as indicated in *Protocol 8*. At the end of electrophoresis the gel was immersed in a solution of ethidium bromide (1 mg/500 ml water, taking care over the hazardous nature of ethidium bromide and its disposal). (a) When viewed under the fluorescent light of a transilluminator, defined bands are observed which correspond to the 28S and 18S ribosomal RNA species. The lack of such bands and an ill-defined streakiness throughout the lane is an indication of RNA degradation.

Subsequent hybridization with a specific probe for a defined mRNA should show a distinct band corresponding to the message species of the expected size. In (b) the mRNA for phosphoenolpyruvate carboxykinase isolated from cultured rat hepatocytes is highlighted by a ^{32}P-labelled cDNA probe. Lane 1 is from cells incubated with 8-(4-chlorophenylthio) cyclic AMP (final concentration, 10^{-6} M) for 2 h. Lane 2 shows the selective decrease in this mRNA when insulin (final concentration, 10^{-7} M) is added to incubations at the same time as the cyclic AMP analogue.

When one is certain of routine intactness and composition of RNA preparations then dot- and slot-blotting may offer advantages for multiple samples.

With intact RNA and a specific probe it is possible to obtain information about the cellular contents of the mRNA encoded by the gene of interest. By applying equal amounts of total RNA in different wells of the agarose gel and then by subsequent scanning for the amount of probe hybridized it is possible to determine if treatment regimes whch modify expression of a particular protein do so at the level of changes in mRNA amount (*Figure 3*). Differences between samples could result from inaccurate loading of RNA on to gels leading to artefactual results. Examination of ethidium bromide staining gives a guide to loading but accurate calibration can be obtained by determination of the amount, for each sample, of a mRNA which does not respond to the cell treatment conditions, i.e. an internal standard. The mRNA chosen will depend on the cell type and the treatment regime but actin, tubulin, and serum albumin have been utilized.

2.3.2 Transcriptional rate: nuclear run-off assays

Direct estimation of the rate of transcription of specific genes comes from the use of nuclear run-off (or run-on) assays (17, 18). In essence the technique relies on the isolation of intact nuclei from tissue or isolated cells. For any particular gene of interest a certain percentage of the population of nuclei will have RNA polymerase II active in transcription along some part of the length of the gene. The intention with this technique is to utilize the isolated nuclei in the test tube to complete the process of transcription of the gene in the presence of radioactive ribonucleotides. Thus transcript amount can be assessed from the extent of radioactive material produced which, in turn, can undergo hybridization to specific nucleic acid probes. The technique should be viewed as having limitations. There is no re-initiation of gene transcription to significant level. There are several reported methods for the production of nuclei for analysis of transcription (17, 18). Variations on the use of detergents and homogenization abound but it is clear that the quality of nuclei produced is the key factor in the determination of transcriptional rate. However, this technique has been extensively used to compare transcriptional rates of nuclei isolated from cells which have been incubated under conditions which modify mRNA amount. As described for measurement of mRNA amount, the transcription rate for a gene which does not alter in response to external conditions should be assessed, in parallel, as an internal standard. The data obtained can indicate whether changes in mRNA content are a consequence of changes to transcriptional events or as a result of changes to the rate of mRNA degradation.

3. Methods for enhancement of protein expression

To give a universal statement on the means for increasing protein expression is not readily feasible. The controlling or limiting factors may be different for each cell type and they may also differ for two proteins within one cell type. Analysis of the levels of protein expression outlined above may help to identify the limiting events. It is clear that protein expression could be modulated in the two following areas:

- at the level of control of gene expression
- at the level of the translational and post-translational apparatus of the cell

In the case of recombinant cells the nature of the vector used for transfection may be relevant for determination of the level of recombinant gene expression (see Chapter 4 and references 19, and 20). For example, the presence of a selectable marker gene such as dihydrofolate reductase in mutant Chinese hamster ovary cells allows amplification of the gene of interest. The presence of several copies of the recombinant gene would then favour formation of the desired protein. However the localization of the recombinant gene must be favourable for expression. The inclusion in the vector of a strong promoter will favour transcription. Construction of the vector may also permit the inclusion of:

- inducible promoters, e.g. mouse mammary tumour virus
- sequences which confer stability to RNA transcripts
- sequences which enhance the interaction of RNA with ribosomal initiation complexes

As yet, there is no detailed description of the ideal vector to permit high levels of recombinant gene expression. The presence in cells of tissue-specific regulators of gene transcription may restrict the ability to define a universal 'supervector'.

Other parameters are relevant to both recombinant and endogenous genes. The structural state of the chromatin surrounding a gene may be crucial in determination of transcriptional level. Modification of the genomic environment causes dramatic changes to the transcription of specific genes (21). For example, 5′-azacytidine, an inhibitor of DNA methyltransferase, influences the transcription rate of several genes. Similar observations have been made with other small-molecular-weight materials, e.g. butyrate (which produces acetylation of nuclear proteins). The cellular effects of certain hormones, growth factors, and mitogens are produced at transcriptional level (22). Increased understanding of the mechanisms by which these actions are provoked at the level of chromatin structure or control of DNA-binding protein function will open new approaches for engineering novel vectors and for design of stimulatory regimes to maximize transcription of selected genes.

Limitations at the level of protein translation and processing may be more apparent with increased transcription. A clear balance must exist between all the steps required for eventual protein delivery to the correct place in the cell. It is clear that hormones, e.g. insulin and certain amino acids, can have actions on the level of translation occurring within several cells types (23). Generally, these actions are expressed at the level of the ribosome and it is unclear if selection can be exerted for specific mRNA species. Little information is available about controlling factors which influence the nature of post-translational processing; the selectivity of such events is crucial in formation of the desired cellular product. It is clear that, although we know a great deal about cells in culture, there are still aspects relatively unexplored. This offers fruitful topics for the future. The methods outlined in this review will make it possible for an investigator to examine the contribution of individual aspects involved in the processing of specific proteins and to formulate approaches which may allow increased formation of their product.

Acknowledgements

I am grateful to Alison Bate, Atika Belbrahem, Marcus Hamer, and Andrew Whatmore for their help in the preparation of this manuscript. I also thank the Medical Research Council and the Science and Engineering Research Council for the financial support which has permitted me to accumulate the research expertise outlined in this chapter.

References

1. Lewis, J. A. (1987). In *Lymphokines and Interferons: A Practical Approach* (ed. M. J. Clemens, A. G. Morris, and A. H. Gearing), p. 73. IRL Press, Oxford.
2. Green, A. K., Cotton, R. G. H., Jennings, I., and Fisher, M. J. (1990). *Biochem. J.* **265,** 563.
3. Rosner, M. R. (1986). In *Mammalian Cell Technology* (ed. W. G. Thilly), p. 63. Butterworths, Boston.
4. Low, M. G. and Saltiel, A. R. (1988). *Science* **239,** 268.
5. Drechou, A., Rouzeau, J.-D., Feger, J., and Durand, F. (1989). *Biochem. J.* **263,** 961.
6. Smith, P. L., Kaetzel, D., Nilson, J., and Baezigl, J. V. (1990). *J. Biol. Chem.* **265,** 874.
7. Bulleid, N. J., Curling, E., Freedman, R. B., and Jenkins, N. (1990). *Biochem. J.* **268,** 777.
8. Rine, J. and Kim, S.-H. (1990). *New Biologist* **2,** 219.
9. Chojkier, M., Brenner, D. A., and Leffert, H. L. (1989). *J. Biol. Chem.* **264,** 9583.
10. Ballard, F. J. (1979). In *Techniques in Metabolic Research,* Vol. B210 (ed. H. L. Kornberg, J. C. Metcalfe, D. H. Northcote, C. I. Pogson, and K. F. Tipton), p. 1. Elsevier, Amsterdam.
11. Yamashita, K., Koide, N., Endo, T., Iwaki, Y., and Kobata, A. (1989). *J. Biol. Chem.* **264,** 2415.
12. Parekh, R. B., Dwek, R. A., Thomas, J. R., Rademacher, T. W., Wittwer, A. J., Howard, S. C., Nelson, R., Siegel, N. R., Jennings, M. G., Harakas, N. K., and Feder, J. (1989). *Biochemistry (Washington)* **28,** 7644.
13. Parekh, R. B., Dwek, R. A., Rudd, P. M., Thomas, J. R., Rademacher, T. W., Warren, T., Wun, T. C., Hebert, B., Reitz, B., Palmier, M., Ramabhadran, T., and Teimeier, D. C. (1989). *Biochemistry (Washington)* **28,** 7670.
14. Eisenstein, R. S. and Harper, R. E. (1984). *J. Biol. Chem.* **259,** 9922.

15. Chirgwin, J. M., Przybyla, A. E., MacDonald, R. J., and Rutter, W. J. (1979). *Biochemistry (Washington)* **18,** 5294.
16. Sambrook, J., Fritsch, E. F., and Maniatis, T. (ed.) (1989). *Molecular Cloning. A Laboratory Manual* (2nd edn). Cold Spring Harbour Press, Cold Spring Harbour, New York.
17. Ausubel, F. M., Brent, R., Kingston, R. E., Moor, D. D., Seidman, J. G., Smith, J. A., and Struhl, K. (ed.) (1987). *Current Protocols in Molecular Biology.* Wiley, New York.
18. Berger, S. L. and Kimmel, A. R. (ed.) (1987). *Methods in Enzymology*, Vol. 152. Academic Press, New York.
19. Weymouth, L. A. and Barsoum, J. (1986). In *Mammalian Cell Technology* (ed. W. G. Thilly), p. 2. Butterworths, Boston.
20. Shyu, A.-B., Greenberg, M. E., and Belasco, J. G. (1989). *Genes Develop.* **3,** 60.
21. Harris, M. (1982). *Cell* **29,** 483.
22. Granner, D. K. and Pilkis, S. J. (1990). *J. Biol. Chem.* **265,** 10173.
23. Olivier, A. R., Ballou, L. M., and Thomas, G. (1988). *Proc. natl Acad. Sci. USA* **85,** 4720.

6

Hybridomas: production and selection

COLIN HARBOUR and ANNE FLETCHER

1. Introduction

1.1 Aims and scope

The first monoclonal antibody (mAb) that we produced immobilized a motile suspension of the bacterium *Salmonella typhimurium*. It was an exhilarating moment when, almost immediately following the addition of our mouse mAb to a highly motile culture of bacteria, we witnessed the rapid cessation of motility. Ten years on, and after the production of hundreds of different mAbs, it is still an exciting moment when a new antibody specificity is detected. The major aim of this chapter is to ensure that the reader can share in a similar feeling of genuine achievement. The generation of mAbs can be an expensive, extremely labour-intensive, and often frustrating task but, providing the immunized mouse has produced antibodies to the chosen antigen, mAbs should be generated using the protocols we describe. We have found, however, that the production of stable cell lines producing human mAbs is much more difficult than in the mouse system. Thus, although the protocols described have proved successful, it is much more difficult to succeed in the production of human rather than mouse mAbs.

In this chapter we describe detailed protocols for the generation of both mouse and human mAbs and for the various assay systems which can be used for detecting the antibodies. In addition we describe methods for maintaining the hybridoma cells and discuss some of the methods available for producing large quantities of antibodies. We also aim to provide the reader with a brief theoretical background in order to understand the basic mechanisms involved in hybridoma technology.

1.2 Background

Although a detailed description of the regulation of antibody formation is beyond the scope of this chapter it is necessary to be conversant with the basic concepts in order to understand the process known as hybridoma technology. Antibodies are glycoproteins of animal origin which provide humoral immunity and form a major part of the immunological defences against infection by invading micro-organisms. Since their discovery they have been widely used as diagnostic reagents and, more recently, as therapeutic biologicals. In the production of diagnostics, appropriate animals, such as rabbits, sheep, or goats, are immunized with the antigen of choice and then bled to obtain sera containing the required antibodies. These sera are known as polyclonal sera since they contain a mixture of antibodies with different specificities directed at the antigen of interest. Antibodies are synthesized by the B lymphocytes although other cells, such as T cells and macrophages, are involved in the complex regulation of antibody formation. The clonal selection theory (1) proposed that each B lymphocyte had a unique receptor specificity for antigen and was predetermined to synthesize only one antibody after the appropriate antigen stimulation. Thus a population or clone of cells derived from a single B cell should secrete antibodies with identical specificity, i.e. mAbs. Unfortunately, this property could not be exploited *in vitro* because B cells are short-lived in cell culture.

Immunoglobulin-secreting cells or myelomas induced chemically in mice or occurring naturally were used as a source of monoclonal antibodies but were of no practical use because they were generally of ill-defined specificity. However, it was shown that fusion of two immunoglobulin secreting cell lines *in vitro* led to the co-dominant expression of the immunoglobulin chains (2, 3). Following these studies Köhler and Milstein (4) made their momentous discovery whereby they fused myeloma cells with splenic lymphocytes from a mouse immunized with sheep red blood cells and showed that some of the resulting hybrid cell lines secreted antibody specific for the sheep red blood cells. Thus by cell fusion it was now possible to generate antibodies of predetermined specificity, uncontaminated with other antibodies, from cells which had the property of continuous growth.

mAbs clearly have a number of advantages over polyclonal sera including:

- unique specificity, unlike polyclonal sera which contain a mixture of specificities
- no batch to batch variation since mAbs are produced from identical cells all derived from a single B cell clone
- unlimited quantities can be produced because hybridoma cells can grow continuously in culture

Thus, mAbs have replaced polyclonal sera in many areas of immunodiagnostics and are being increasingly used on a large scale for many other applications including: the purification of biologicals, e.g. blood products such as the coagulation protein, factor VIII; in immunoimaging and immunotherapy of cancers, that is the detection of cancer cells in the body using radiolabelled antibody and the destruction of such cells with antibody–toxin conjugates; in conventional immunoglobulin therapy for the prevention of infections; and in the prevention of graft rejection in kidney transplant recipients (5).

2. Theory

In this section we describe the theoretical background underpinning hybridoma technology while also presenting an alternative method for generating human mAbs, that is Epstein–Barr virus (EBV) mediated transformation of lymphocytes. The essential ingredients for the successful production of a specific mAb by fusion technology include a B cell population stimulated by the appropriate antigen and a suitable cell line to act as a fusion partner.

2.1 Fusion partners

Some of the cell lines available as fusion partners are shown in *Table 1*. It is highly desirable that the fusion partner does not synthesize antibody; otherwise the resulting hybrid could secrete a mixture of antibodies with no or reduced specific activity. As the table shows, several such non-secreting cell lines are available. We have successfully used both P3-NS1/1-Ag4-1, abbreviated to NS1, and P3-X63-Ag8-653, abbreviated to X653, for the production of mouse mAbs, although the latter is preferred since it does not synthesize antibody.

2.2 Selection of hybrids

The fusion will result in a heterogeneous cell population containing the following possibilities: lymphocyte × lymphocyte hybrids; lymphocyte × myeloma hybrids; myeloma × myeloma hybrids; unfused myelomas and un-

Table 1. Immortal cell lines used as fusion partners

Species	Cell line	Cell phenotype	Ig expression	Reference number
Human	SK007	Myeloma	IgE(λ)	6
	RH-L4	Lymphoblastoid	$IgG(k)_1$ non-secretor	7
	GM1500	Lymphoblastoid	IgG(*k*)	8
	KR4	Lymphoblastoid	IgG2(*k*)	9
	LICR-LON HMy2 }	Lymphoblastoid	IgG1(*k*)	10
Human/human	KR12	Hybrid myeloma	IgG(*k*)	11
Human/mouse	SHM-D33	Hybrid myeloma	IgE non-secretor	12
Mouse	X653	Myeloma	None	13
	NS0	Myeloma	None	14
	Sp2/0	Myeloma	None	15
	NS1	Myeloma	IgG1(*k*)	16
Rat	Y3-Ag 1.2.3.	Myeloma	*k*	17
	YB2/0	Myeloma	None	18

fused lymphocytes. The lymphocyte hybrids and unfused lymphocytes will quickly die but the myeloma cell lines used as fusion partners will continue to grow, and probably outgrow, the desired hybrids. Thus there has to be some form of selection in order to inhibit the growth of the myeloma cells. The murine myelomas have been grown in the presence of either 8-azaguanine or 6-thioguanine which selects for the growth of variants defective in the enzyme hypoxanthine–guanine phosphoribosyl transferase (HGPRT). This enzyme is involved in a salvage pathway for DNA synthesis as shown in *Figure 1*. If aminopterin, which blocks the main DNA synthetic pathway, is included in the medium then the HGPRT$^-$ myeloma cells will die (19). However, hybrid cells formed by the fusion of HGPRT$^-$ myeloma cells and spleen cells will survive in the presence of aminopterin, providing they have inherited a functional HGPRT enzyme from the spleen cells. In order that the salvage pathway functions effectively it is necessary to add exogenous nucleotides, i.e. hypoxanthine and thymidine.

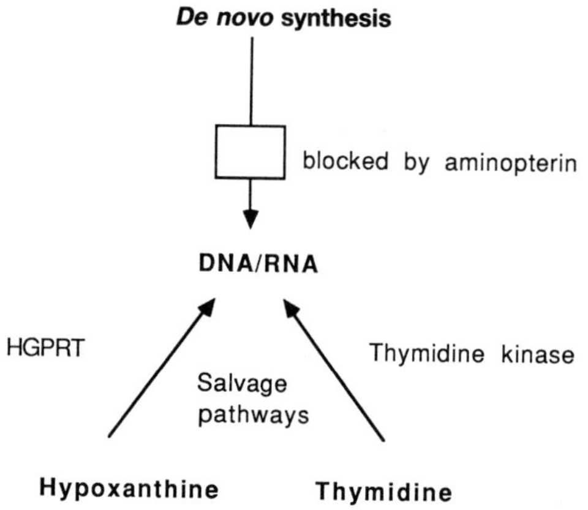

Figure 1. HAT selection of hybridoma cells.

2.3 Fusion process

In the fusion process myeloma and lymphocyte cells are brought close together by centrifugation in the presence of the chemical polyethylene glycol (PEG) which results in the fusion of adjacent cell membranes and hence the formation of hybrid cells. In the original report (4) Sendai virus was used as the fusing agent but this is now not routinely used. More recently, the use of electrical pulses has been employed successfully to form hybrids by the process known as electrofusion (20). A summary of the various steps involved in mAb production is shown in *Figure 2*. Each step is discussed in detail in Sections 6 and 7. The fusion protocol that we describe is based on the procedures described by several groups of workers including Galfrè and Milstein

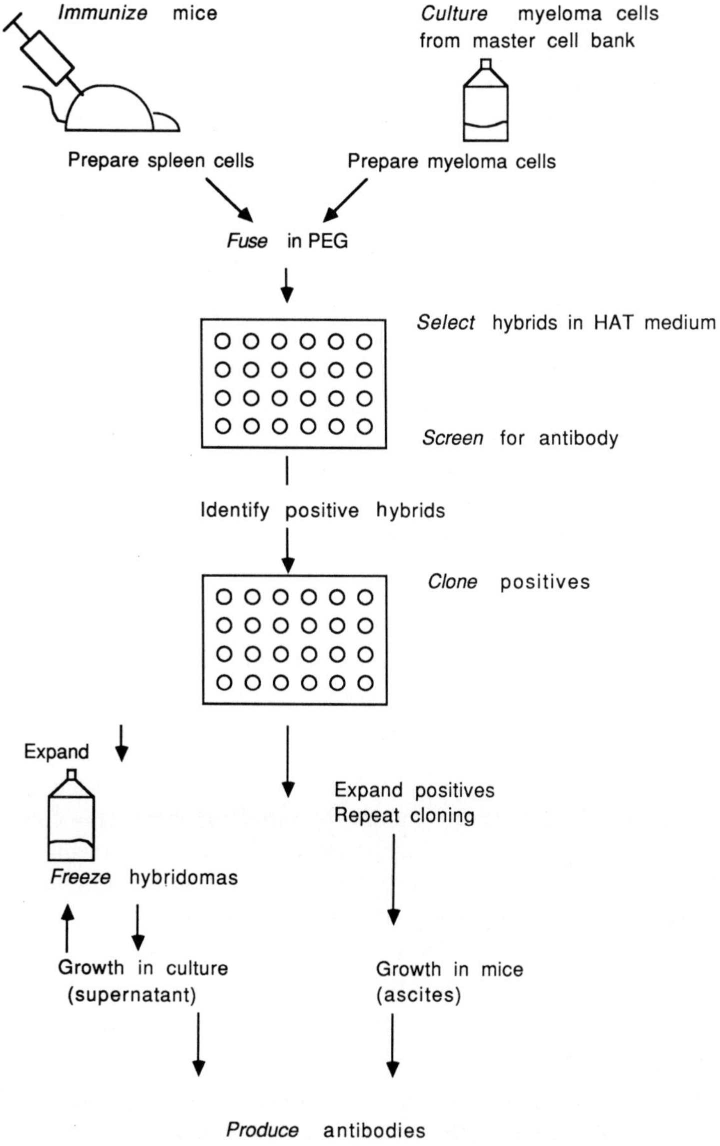

Figure 2. Steps in monoclonal antibody production.

(14), Oi and Herzenberg (21), and Zola and Brooks (22). In general, fusion technology works extremely well for the generation of mouse mAbs but less well for human mAbs for the reasons outlined in the following section.

2.4 Human monoclonal antibodies

Human mAbs will be required in instances where rodents are unable to produce human antibody specificities such as in the case of the human erythrocyte antigen, Rh(D), and to a large extent for the human leucocyte (HLA) antigens. In addition, human antibodies are preferred for therapy to rodent antibodies because of the risk of eliciting an anti-rodent immune response and for studying the human immune system. The technology of human mAb production has proved difficult and at present there is no universally reliable and accepted method for routine use. The various strategies which have been employed with differing success rates have been recently reviewed (5, 23).

There are two major difficulties in generating human mAbs. First, it is difficult to obtain a rich source of immune human B cells. This is because the most readily available source of B cells is peripheral blood since spleen and lymph nodes are only rarely obtainable and yet specific antibody-producing cells are infrequent in peripheral blood. Moreover, B cells from blood may not be in an appropriate state of differentiation and activation for successful fusion and immortalization (24). In many instances it is impossible to immunize human donors for the antigen of interest. Attempts have been made to increase the number of specific antibody-secreting cells by specific enrichment, *in vitro* stimulation, or polyclonal activation (5). The absence of a satisfactory human cell line for use as a fusion partner is the other major difficulty. Human myeloma cell lines are very difficult to grow in culture and have a low fusion efficiency. Other human cell lines such as lymphoblastoid lines may be grown more easily but are less well adapted for high levels of immunoglobulin secretion (25). Unlike the widely used non-secreting mouse myeloma cell lines, most of the human cell lines used as fusion partners synthesize and secrete human immunoglobulins. This complicates the characterization of hybrid immunoglobulins and may result in mixed molecules (26). When compared with murine hybrid production fusion yields are lower and human hybrids are more often unstable.

The major approaches to generating human mAbs are:

- hybridization of lymphocytes with a fusion partner
- immortalization of lymphocytes by Epstein–Barr virus (EBV) transformation
- a combination of the above techniques in which EBV immortalized cells are fused with a myeloma or hybrid myeloma fusion partner

A large number of cell lines has been investigated for suitability as fusion partners with human B cells. These include human lines, mouse–human heteromyeloma lines, and mouse myeloma lines. Examples of the cell lines used are given in *Table 1*. Immortalization of lymphocytes by EBV transformation avoids the requirement for fusion; however, it has been generally reported that the lymphoblastoid lines which result can only be cloned with difficulty and may undergo senescence (27). Others have succeeded in obtaining stable lines (28, 29). It is possible to obtain clonal cultures from the outset by adoption of limiting dilution techniques (30). Enrichment of antibody-secreting cells by procedures such as rosetting may assist in obtaining stable lines (31). In order to overcome the disadvantages associated with each of the above approaches a number of workers have first expanded the antibody-secreting cells of interest by EBV transformation and then fused and obtained hybrids. This requires an additional selection procedure to be included in the fusion step since the EBV-immortalized cells need to be selected against. This is most often achieved by fusion with an ouabain-resistant fusion partner and then selection of hybrids in ouabain and HAT. A selection process for obtaining fusion partners with ouabain resistance is described by Kozbor *et al.* (9).

3. Electrofusion

This is a technique which has been used successfully for producing both mouse and human mAbs (19, 32). Electrofusion is based on the fact that the cell membrane is made permeable for a short period in response to electrical breakdown. This phenomenon was first described in 1973 (33) and has been used for the electro-injection of membrane-impermeable substances, such as DNA, into freely suspended cells without adverse affects. We do not describe this method in this chapter but for those interested we refer the reader to the work of Zimmermann and colleagues (20).

4. Immunization procedures

The generation of mAbs involves at some stage experimentation with either animal species, usually mice, and/or humans. There are, therefore, ethical and moral issues involved in such experiments which must be considered. In recent years the use of animals has become more closely regulated and each institution involved in animal experimentation has to follow legislative procedures. For example, an institution has to have an Animal Ethical Review Committee to which all projects involving the use of animals must be submitted for approval. The committee should include non-scientists from the staff of the institution and members of the public. Thus the first step in producing a mAb is to submit the proposal to the Animal Ethical Review Committee. The generation of mAbs involves the use of lymphocytes derived from a donor or patient. It is clearly essential to obtain the donor's consent before starting the experiment although the full ramifications of what happens if an antibody is produced and used commercially are probably yet to be resolved.

The immunization strategy that is adopted will depend on a number of factors including: (a) whether mouse, rat, or human mAbs are desired; (b) whether a particular class or subclass of antibody is required; and (c) the type, amount, and purity of antigen available.

4.1 Source of stimulated B cells

For the production of mouse mAbs, Balb/C mice are most frequently used because the mouse myeloma cell lines that are available for hybridoma formation are of Balb/C origin. Other mice can be used successfully; for example, we have used C57B1 mice to produce mAbs but the resulting hybridoma cells cannot be grown in these mice. In recent years, with the development of therapeutic uses for mAbs and the recognition that laboratory rodents may harbour microbial agents infectious for man, the quality control of animals has improved significantly. It is recommended that specific pathogen-free mice should be used and that the colony be screened regularly for infectious agents.

It should be possible to generate mouse mAbs to any antigen, providing the antigen is not of murine origin. There are some exceptions; for example, despite numerous attempts it has not proved possible to produce mouse mAbs specific for the Rh(D) blood group antigen. In this case human mAbs must be produced, although our success rate is significantly lower than with the mouse system. The lymphocyte donors would have to have a high level of the desired antibody in their blood since it may be neither practical nor ethical to boost human volunteers prior to obtaining B cells for immortalization. Due to the difficulties of immunizing human donors, *in vitro* stimulation strategies are particularly important and these are discussed in Section 4.3.

4.2 *In vivo* immunization strategies

Mice can be immunized subcutaneously, intramuscularly, intraperitoneally (i.p.), intravenously (i.v.), or intrasplenically (34). The most common method is i.p. For successful immunizations soluble antigens usually require the use of an adjuvant, such as Freund's complete adjuvant (FCA), whereas cells, such as lymphocytes, red cells, and bacteria, can be injected without the use of an adjuvant. A simple immunization schedule would be as follows:

- day 1: first injection of 6–8 week old Balb/C mice
- day 28: second injection
- day 31: remove spleen cells for fusion

4.2.1 Generation of anti-bacterial antibodies

We have successfully used this simple immunization schedule to generate mAbs specific for the flagella antigens of *Salmonella typhimurium*. The bacteria were injected i.p. in 0.3 ml volumes containing 10^8 bacteria/ml. The bacteria were prepared for injection by centrifuging the bacterial culture broth, washing the cell pellet three times with phosphate-buffered saline (PBS), fixing with 0.1% (v/v) formalin for 1 h, and then re-washing with PBS before final re-suspension to the desired cell concentration. Using the stated method we obtained immunoglobulin IgG2a antibodies which agglutinated the bacteria. We were also successful in producing mAbs of similar specificity but of IgM class using the following protocol:

- day 1: first injection
- day 5: removal of spleen cells for fusion

In general, it is not as simple as this and the immunization protocol must be optimized by using different doses of antigen, with and without adjuvant, trying a range of booster injections, and varying the time interval between the booster injections. Although it is usual to carry out the fusion 3–5 days after the final booster, we have successfully employed the technique of Stahli *et al.* (35) for the generation of mAbs specific for red blood cell antigens in which five booster injections are given in the week before fusion. This procedure is recommended for antigens which are weakly immunogenic. The antibody response is monitored by bleeding the mice at regular intervals. If the antibody titre is satisfactory then the final booster can be given either i.p. or i.v.

4.2.2 Generation of antibodies to soluble antigens

Soluble antigens are injected in adjuvant at concentrations in the range 5–100 μg/mouse although the concentration is obviously dependent on the purity of the antigen preparation. The antigen is thoroughly emulsified in FCA in a 1:1 ratio (great care should be taken in handling FCA) and 0.2 ml injected i.p. or in several sites intradermally. The booster injections, usually two or three at 2–4 week intervals, containing the same dose of antigen, are given in incomplete Freund's adjuvant to avoid the development of hypersensitivity reactions. The final injection may be given without adjuvant. For antigens which either do not generate an immune response using adjuvant or are in limited supply, *in vitro* immunization may be appropriate. As an example of a soluble antigen we have successfully produced mAbs to the hepatitis B virus surface antigen.

Protocol 1. Immunization protocol for soluble antigens

1. Dilute highly purified surface antigen in saline and emulsify 1:1 with FCA.
2. Inject 0.3 ml of the mixture containing approximately 40 μg of the antigen i.p.
3. Five to six weeks later dilute the same antigen in incomplete Freund's adjuvant and inject 0.3 ml containing 80 μg i.p.
4. Perform the fusion 3 days later (see *Protocol 7*).

From the immunization scheme described in *Protocol 1*, 150 of the 384 microcultures established from the fusion showed hybridoma growth and 50% of the 150 cultures were found to produce antibodies to the surface antigen. In a parallel fusion carried out using a mouse of the same age and using the same antigen concentrations, but without using FCA in the primary injection, 82 of 384 microcultures showed hybridoma growth but none of these cultures secreted specific antibody. This finding underlines the importance of using an adjuvant with soluble antigens and the need to optimize the immunization schedule.

4.2.3 Generation of antibodies to animal cells

In our experience it is not easy to generate antibodies of blood group specificity, e.g. anti-B, but we have obtained agglutinating anti-red blood cell antibodies by injection 0.1 ml of a 50% solution of washed red blood cells in PBS, i.e. 5×10^8 cells, on each of days 1 and 14 and then fusing on day 17. Alternatively, the injections can be given a month apart.

4.3 *In vitro* immunization

This system is particularly useful for the generation of human mAbs where the immunization of human volunteers to obtain stimulated B cells may not be appropriate. The most suitable sources of B cells are the spleen and lymph nodes which are only rarely obtainable from human sources. The most readily available source of B cells is peripheral blood in which the supply of specific antibody-producing cells is infrequent. Moreover, B cells from the blood may not be in an appropriate state of differentiation and activation for successful fusion and immortalization (24). Thus attempts have been made to increase the number of specific antibody-secreting cells by specific enrichment or *in vitro* stimulation. This is now an area of intensive research and several methods have been published which should be referred to for more detail (36–38). We recommend that *in vitro* immunization be used prior to mouse and human fusions but not before EBV transformation because lymphocytes lose the EBV (CR2) receptor during stimulation. A detailed protocol is provided by Newell *et al.* (39) for the stimulation of mouse B cells *in vitro*.

5. Selection procedures for monoclonal antibodies

It is somewhat paradoxical that, while the majority of mAbs are produced specifically for use in a particular assay, it is essential to develop an assay to detect the mAbs. This section on selection and screening is placed prior to the fusion process because the selection of an assay is extremely important. A specific, sensitive, reliable, rapid, and reasonably economic assay has to be developed long before the fusions are commenced. There are a number of assays which meet these criteria; these include:

- enzyme-linked immunosorbent assay (ELISA)
- agglutination or haemagglutination assays

- radioimmunoassays
- immunofluorescence tests
- immunoblot assays

The choice of an assay system will depend upon the use intended for the mAb. For example, if the mAb is being produced for the development of an ELISA assay, then this assay will have to be used in the screening process since an antibody which reacts in an ELISA may not agglutinate or precipitate and vice versa. The ELISA (40) is now the most popular assay for screening of hybridoma supernatants. It can be used for detecting antibodies to both soluble antigens and whole cells. In the ELISA the antigen is bound on to the surface of a 96-well microtitre plate. It is important to establish the optimum antigen concentration for use and this is usually done by performing a chequerboard titration in which different concentrations of antigen are tested against different concentrations of positive and negative polyclonal sera from immunized and non-immunized mice respectively. Using the optimum antigen concentration, further tests should be carried out to determine the appropriate concentration of anti-mouse antibody–enzyme conjugate for use in the assay. Although appropriate dilutions are recommended by the manufacturers, it is useful to determine the optimum dilution for each ELISA assay.

Protocol 2. A typical ELISA assay

1. Prepare the following reagents:
- Coating buffer (80 ml of 0.2 M Na_2CO_3, 220 ml of 0.2 M $NaHCO_3$). Dilute to 1 litre with distilled water and adjust pH to 9.6.
- Phosphate buffered saline (PBS) (8.5 g NaCl, 1.07 g Na_2HPO_4, 0.39 g NaH_2PO_4). Dissolve in distilled water and make up to 1 litre.
- Diluent and blocking buffer (1% bovine serum albumin in PBS or 0.1–1% skim milk in PBS). The latter is usually filtered before use to clarify
- Washing buffer. Make up 0.05% Tween 20 in PBS
- Enzyme conjugate. Purchase affinity purified peroxidase or alkaline phosphatase labelled goat anti-mouse immunoglobulin (Ig). Reconstitute in 1 ml of 50% aqueous glycerol to give a final concentration of 0.1 mg/ml. Store the conjugate at −20°C. Dilute the conjugate with diluting buffer by an amount determined by the preliminary optimization studies (approximately 400 times).
- Enzyme substrate. Several alternatives are available but we routinely use 2,2′-azino-di-(3-ethylbenzathiazoline) sulphonate (ABTS) and hydrogen peroxide mixed 1:1 immediately before use with peroxidase conjugates. Phosphatase substrate (Sigma 104) may be used with the

Protocol 2. *continued*

alkaline phosphatase conjugate at 10 mg/ml. The phosphatase substrate is dissolved in 0.5 M glycine buffer pH 10.4 containing 5 mM $ZnCl_2$ and 5 mM $MgCl_2$.

2. Add dilutions of the antigen in coating buffer (100 μl) to each well of a 96-well microtitre plate suitable for ELISA assays. The antigen concentration is dependent on the preliminary optimization studies but is usually 1–10 μg/ml for soluble antigens and approximately 10^8–10^9 cells/ml when detecting anti-bacterial antibodies. Purified protein antigens can be used directly but low-molecular-weight antigens require complexing with a large molecule such as bovine serum albumin.
3. Incubate the plate overnight at 4°C or at room temperature in a moist container.
4. Rinse plates with 200 μl wash buffer per well. Shake out and blot dry. Repeat three times.
5. Add 100 μl of blocking buffer to each well and leave for 1 h at 37°C.
6. Wash plates as before three times.
7. Add 100 μl aliquots of antibody standards or samples of culture supernatant to each well. Include various dilutions and appropriate negative and positive controls.
8. Incubate the plate at 37°C for 1 h in a humid container.
9. Wash three times (as before) and repeat using PBS alone.
10. Add 100 μl of diluted enzyme conjugate to each well. Leave for 1 h at 37°C.
11. Wash three times with wash buffer and then three times with distilled water.
12. Add substrate (ABTS is prepared immediately before use) and read optical density at 405 nm at 5 min intervals from 0 time to 20 min. Alternatively, the reaction can be stopped at a specified time (e.g. 1 h).

6. Production of mouse monoclonal antibodies

6.1 Equipment

Most tissue culture laboratories will already possess all the equipment necessary for the generation of mAbs. It is important to realize, however, that mAb production is very demanding of time and it is useful to have a dedicated laminar air flow cabinet and CO_2 cabinet. If other workers are growing non-hybridoma cell lines and have to use the same cabinet, then it would be sensible practice to ensure that the cabinet is thoroughly disinfected and left running for 10 to 15 min before any hybridoma culture work is begun. A similar rule should apply when you are working with more than one hybridoma cell line so as to avoid cross-contamination. If both murine and human mAbs are being produced in the same laboratory then it is important to ensure that the cell lines are kept separate. Ideally, this should involve the use of separate laminar air flow cabinets and CO_2 cabinets. It is important to note that if, at a future date, the cell line and its product is considered suitable for therapeutic application then a detailed history will have to be provided to the licensing authorities. The latter would be concerned if there was a risk of cross-contamination between human and murine cell lines. A detailed history of the hybridoma cell line should be kept.

6.2 Media and reagents

6.2.1 Culture media

Although RPMI 1640 is usually the medium of choice when working with lymphocyte cultures, Dulbecco's modified Eagle's medium (DMEM) has proved equally as suitable for use with murine hybridoma cells. Other media, such as Iscove's modified DMEM (IMDM), can also be used for hybridoma cultivation (see Chapter 1).

6.2.2 Fetal calf serum (FCS)

FCS is an essential additive for media used for fusions although, once the hybridoma cells are established in cultures as cell lines, it may be possible to wean the cells into low-serum or serum-free media. Prior to the purchase of the FCS batch it is advisable to test several batches for their ability to support the growth of the myeloma cell line. Following purchase, the FCS should be stored at −20°C to −70°C until required. Before use, thaw at room temperature and heat inactivate at 56°C. The length of time required for inactivation depends on the volume of FCS, e.g. we use 1 h for 100 ml bottles and 2 h for 500 ml. The serum is mixed during this period. The FCS is then stored at 4°C before use. FCS is used at 10–20% (v/v) for fusion media but when hybridoma cell lines are established we routinely reduce FCS to 5% (v/v).

6.2.3 Polyethylene glycol

Polyethylene glycol (PEG) is toxic to cells and some batches are more toxic than others. It is therefore important to carry out fusion trials to establish that the batch and molecular weight, e.g. 4000 or 1500 PEG, is suitable. The PEG solution is prepared as in *Protocol 3*.

Protocol 3. Preparation of PEG solution

1. Weigh 30 g PEG 4000 (BDH) into a 100 ml sterile glass bottle.
2. Autoclave for 20 min at 121°C to dissolve the PEG.

Protocol 3. *continued*

3. Add 40 ml of a solution containing 15% dimethylsulphoxide (DMSO) in Dulbecco's phosphate buffered saline (PBS), e.g. 6 ml DMSO + 32 ml PBS.
4. Mix thoroughly and distribute 10 ml aliquots to McCartney bottles.
5. Autoclave at 121 °C for 30 min and store at 37 °C. It is important not to allow the PEG to cool below 37 °C so as to avoid precipitation. The PEG should remain stable for approximately 6 months.

6.2.4 Hypoxanthine, aminopterin, and thymidine (HAT)

It is possible to obtain these chemicals in powder form separately in order to prepare the mixtures but since the aminopterin is reportedly a teratogen we minimize handling by using ready-made formulations. We purchase 50× strength solutions (e.g. Flow Laboratories) of both HAT and HT, which are supplied frozen, and store at −20°C. When required, the stock solution is thawed, 2 ml aliquots distributed into sterile bijou bottles, and frozen at −20°C. The final concentrations required in the medium are 10^{-4} M hypoxanthine, 1.6×10^{-5} M thymidine, and 4×10^{-7} M aminopterin.

6.2.5 Sucrose

Sucrose is used to obtain feeder cells. A 0.34 M solution is prepared, usually in 100 ml aliquots, and then sterilized by autoclaving for 15–20 min at 115°C. Care should be taken to avoid the temperature going above 115°C to avoid caramelization of the sugar.

6.2.6 Feeder cell layer

The majority of hybridoma cells when inoculated at low cell density will not grow in the absence of feeder cells, such as macrophages, which are thought to supply growth factors. As an alternative, growth factor preparations are now available commercially, e.g. BM Condimed HI (Boehringer Mannheim), and these are more convenient to use as 10% v/v supplements but they may not always work effectively in fusions. We tend to use the supplements for cloning cells but have retained the use of feeder cells for fusions.

Protocol 4. Preparation of a macrophage feeder cell layer

1. Kill the mice by CO_2 asphyxiation and pin to a dissection board with ventral surface uppermost. Two mice are usually sufficient for one fusion using 5 × 96-well tissue culture plates.
2. Swab the skin thoroughly with 70% alcohol and place in the laminar air flow cabinet.
3. Tear the ventral skin transversely at the median line to expose the peritoneal membrane.

4. Inject 5 ml of cold (4°C) 0.34 M sucrose into the lower peritoneum using a 21 g needle, being careful not to puncture any internal organs.
5. Wait approximately 1 min and withdraw as much of the sucrose solution as possible. If the solution appears to be pink when withdrawn it will contain mainly red blood cells and should be discarded.
6. Count the number of macrophages obtained using a haemocytometer.
7. Calculate the volume of medium required to re-suspend the cells so as to provide a final concentration of 10^5 cells/ml. The aim is to add approximately 10^4 macrophages to each well of a microtitre plate in 100 μl of medium.
8. Spin the sucrose solution containing the cells at 400 *g* for 10 min.
9. Decant the supernatant and gently re-suspend the cell pellet in the correct volume of DMEM + 10% FCS medium. The macrophages should be used immediately but, if a delay is unavoidable, the preparation should be kept on ice until use to avoid the macrophages attaching to the container being used.
10. Distribute 100 μl of the cell suspension to each well of a 96-well microtitre tissue culture plate. We usually use four plates per fusion. Cover the plate with the lid and seal with tape.
11. Incubate the plate at 37°C in a humidified CO_2 cabinet at least 1 h before adding any fused cells but preferably overnight.

6.3 Maintenance and cultivation of myeloma cells

It is important for the success of the fusion that the myeloma cells are in the mid-log phase of growth and hence actively growing. Also, in order to ensure that the myeloma cell line has grown for approximately the same number of generations prior to each of the fusion experiments, it is advisable to prepare a large stock of frozen vials of the cell line as soon as possible after it is received by the laboratory (see Chapter 1). Then, approximately 7–10 days prior to the planned fusion date, a vial of frozen cells is removed from liquid nitrogen and cultured as described below. Alternatively, if a number of fusions are planned over a short period of time, say no more than 4 weeks, then the myeloma can be maintained in culture by subculturing every Monday, Wednesday, and Friday until the fusion programme is complete. In practice, however, we have found that the NSI-cell line can be cultured for over a year, involving three subcultures per week, and employed in successful fusions. We found the fusion efficiency did not alter significantly during the period of the study. This approach is not recommended, however, because it is both time-consuming and expensive in materials and media to maintain the cell line by repeated subculture and could result in deterioration of the myeloma cell line.

We therefore recommend freezing so as to obtain both a master bank and working bank of myeloma cells. Prior to the preparation of the working bank, the master bank should be checked for the presence of certain viruses, e.g. the Hantaan virus, lymphocytic chorio-meningitis virus, and murine retroviruses, and mycoplasma contamination. Such services are now commercially available.

6.3.1 Freezing of cells

It is important upon receipt of the myeloma cell line to expand as rapidly as possible to a 10 ml culture in a 25 cm^2 flask and then to a 50 ml culture in a 75 cm^2 flask. Depending on the number of vials required (50 is a useful working number to begin with), multiple 75 cm^2 flask cultures or larger volumes in spinner culture flasks (Techne Ltd) can be established. As a rough guide each frozen vial should contain between 2 and 5×10^6 cells and 10 ml of cell culture will generate sufficient cells for one or two vials. The cells are in an extremely fragile state when removed from liquid nitrogen and great care should be used in handling the cells during the thawing procedure.

6.3.2 Culture of myeloma cells for fusions

Before embarking on a series of fusions it is useful practice to check that the myeloma cells are HAT-sensitive due to the lack of the HGPRT enzyme. This is done by cloning the cells (see *Protocol 9*) in the presence of HAT. If the cells survive, then a new supply of myeloma cells should be obtained. Alternatively, the cells can be cloned in the normal cloning medium supplemented with 8-azaguanine (20 μg/ml). Only $HGPRT^-$ cells will survive in this medium. HAT-sensitive cells can then be prepared for the fusion.

Protocol 5. Culture of myeloma cells for fusion

1. Remove a vial of cells from liquid nitrogen, thaw, and establish a culture (see Chapter 1).
2. On the day prior to a proposed fusion experiment, establish a 50 ml culture of myeloma cells at 10^5 cells/ml in a 75 cm^2 tissue culture flask.
3. On the day of the fusion, determine the cell numbers and their viability. Cells must be at least 90% viable for a successful fusion.
4. Remove an aliquot of cells containing 10^7 viable cells from the culture and centrifuge at 400 *g* for 10 min.
5. After centrifugation decant the supernatant and re-suspend the cell pellet in 10 ml of serum-free DMEM.
6. Repeat the centrifugation, decant the supernatant, and then re-suspend the cell pellet in 10 ml of serum-free DMEM.
7. Repeat Step **6** and use the cells as soon as possible in the fusion.

6.4 Preparation of spleen cells

In the fusion process myeloma cells and spleen cells are mixed in a ratio of approximately 1:10. Some protocols include the counting of spleen cells to ensure that the ratio is within the stated limit, the enrichment of B cells, and lysing red blood cells. We do not routinely employ these procedures for the generation of mouse hybridomas. A single cell suspension of spleen cells is required and it is usual to recover 10^7–10^8 viable lymphocytes from the spleen using the procedure in *Protocol 6*.

Protocol 6. Preparation of spleen cells for fusion

1. Anaesthetize the immunized mouse with diethyl ether 3 to 4 days following the final antigen injection.
2. Pin the mouse to a dissection board with ventral surface uppermost and thoroughly swab with 70% alcohol.
3. Bleed the mouse by either cardiac puncture or via the axilla and retain serum for antibody screen.
4. Kill the mouse by CO_2 asphyxiation, re-swab with alcohol, and place in the laminar air flow cabinet.
5. Tear the ventral skin transversely at the median line using artery forceps to expose the peritoneal wall.
6. Open the peritoneum and remove the spleen using sterile scissors and forceps being careful not to puncture any other organs. Clean off as much fatty tissue as possible.
7. Place the spleen into a sterile Petri dish containing 10 ml of DMEM and gently agitate the spleen in the medium.
8. Transfer the spleen to a second Petri dish containing 10 ml of DMEM and gently tease cells from the spleen capsule using two sterile curved forceps.
9. Remove the cells from the Petri dish using a 10 ml syringe and transfer to a sterile universal container. Allow the cell suspension to stand for a few minutes so that large clumps settle to the bottom. The spleen cells are now ready for the fusion.

6.5 Fusion procedure

In this process the myeloma cells and spleen cells are brought close together in the presence of PEG (*Protocol 7*).

Protocol 7. Fusion of myeloma cells with mouse spleen cells

1. Transfer the spleen cells (see *Protocol 6*) with a Pasteur pipette, taking care to avoid the cell clumps, to a 50 ml centrifuge tube containing the myeloma cells (see *Protocol 5*). Centrifuge at 400 *g* for 10 min.
2. Decant the supernatant and tap the tube gently to re-suspend the cell pellet.
3. Add 1 ml of warm (37 °C) PEG over 1 min using a 1 ml pipette while constantly shaking the cells.
4. Gently centrifuge for 90 sec by increasing the centrifuge speed to 200–300 *g* and then immediately turning the centrifuge off.
5. Add 5 ml of warm (37 °C) serum-free DMEM to the cells gradually using a 5 ml pipette whilst shaking the tube for 1 min (slow addition of warm medium serves to gradually dilute the PEG without lysing the cells).
6. Add 10 ml of warm (37 °C) serum free DMEM over 2–3 min.
7. Centrifuge the suspension at 400 *g* for 10 min at room temperature, decant the supernatant, and tap the tube to re-suspend the cell pellet.
8. Add the cell pellet to 50 ml DMEM containing 10–20% v/v FCS in a 75 cm^2 tissue culture flask. Gas with filtered CO_2/air mixture and incubate at 37 °C for 1 h.
9. After 1 h incubation add 2 ml HAT (50×) and distribute 100 μl of the mixture to each well of five, flat-bottomed, 96-well tissue culture plates containing 100 μl of a macrophage feeder layer (see *Protocol 4*). Seal the plates with tape and incubate in a humidified 37 °C CO_2 incubator. Alternatively, the HAT is added after 12 h.

6.6 Maintenance of fused cells

During the first few days after the fusion the HGPRT-deficient myeloma cells will die in the HAT medium along with the normal spleen lymphocytes. Thus only hybridoma cells will survive and multiply but, since not all hybrid cells will secrete antibody, it is important to monitor the cultures regularly and assay as quickly as possible to avoid the possibility of non-immunoglobulin-secreting hybrid cells overgrowing the antibody secretors. In our laboratories we normally perform the final immunization on a Tuesday, the fusion on a Friday, and then leave the cells untouched until the following Tuesday. The cells are then fed every Tuesday and Friday until the cultures are considered ready for assay. The initial feeding may be delayed to up to 7 days post-fusion without affecting the fusion results.

Protocol 8. Feeding of hybridoma cells

1. A sterile Pasteur pipette attached to a suction device is used to remove half the contents of each culture well.
2. Add 100 μl of fresh medium containing HAT and FCS (10–20%) using a multi-channel pipette and sterile pipette tips. Be careful not to disturb the growing hybridoma colonies.
3. After 10–16 days in culture the HAT supplement is replaced by HT. The HT supplement is continued for at least 7–10 days until all traces of aminopterin have been diluted out of the culture medium.
4. After a few days in culture the cultures are regularly checked to see whether or not hybridoma cells are growing and that the cultures are uncontaminated.
5. Cultures are ready for assay when the cell growth covers about 50% of the bottom of the well and overgrowth is to be avoided. It is important to allow at least 2–4 days after the last feeding before assaying.

6.7 Screening of hybridoma supernatants

50–100 μl of the culture supernatant are removed from the culture well and assayed by one of the procedures described previously. If the culture is positive it is cloned as soon as possible. Cloning may be carried out on the cells from the wells of the 96-well microtitre plate or expanded into a 2 ml culture in a 24-well plate. The latter allows for the freezing, and hence preservation, of the remaining uncloned cells as security in case the cloning is not successful.

6.8 Cloning of hybridomas

The specific aim of cloning is to obtain a population of hybridoma cells derived from a single cell and hence all producing identical antibodies. We clone by limiting dilution and have found that cell lines cloned three times have proved very stable in our hands, apart from the odd exception.

Protocol 9. Cloning of hybrid cell lines

1. Prepare 100 ml of culture medium containing 15–20% FCS (v/v) and either HAT or HT supplement. We seed cloning plates with peritoneal macrophages as feeder cells 24 h before cloning as described in *Protocol 4*. Alternatively, a growth factor supplement can be used.
2. Check the antibody-positive culture well for any contamination or deterioration using the inverted microscope. The cells must be healthy for the cloning step.

Protocol 9. *continued*

3. Gently mix the contents of the culture well to suspend the cells evenly throughout the culture medium. Count the cells and determine viability. The viability should be no less than 80%.
4. Prepare a stock cell culture suspension containing 1000 cells/ml.
5. Dilute the stock solution 20 times with the medium to obtain a cell suspension containing 50 cells/ml and distribute 100 µl of this suspension to each of 48 wells of a 96-well microtitre plate (plate 1). The objective is to obtain approximately five cells per well. *Note*: the wells already contain 100 µl of the macrophage feeder layer.
6. Dilute the 50 cells/ml suspension five times with medium and distribute 100 µl of the resulting cell suspension to each well of a 96-well microtitre plate (plate 2). Thus each well should contain approximately 1 cell per well.
7. Dilute the 50 cells/ml suspension 10 times with cloning medium and pipette 100 µl of the cell suspension to the remaining 48 wells of plate 1.
8. Seal the two microtitre plates with autoclave tape and place in the CO_2 incubator.
9. The plates are regularly examined and fed in the same way as the fusion plates. Assays can usually be performed after 7–14 days. Positive clones are then re-cloned either directly from the microtitre wells or by first expanding the culture into 2 ml cultures and then re-cloning. We usually adopt the latter strategy which allows for the freezing of once-cloned cells and also the expansion of cultures into flasks and the subsequent freezing of more cells. It is important to emphasize that cells must be frozen at each stage, i.e. after the initial assay of the fusion well and after each of the clonings.

7. Production of human monoclonal antibodies

7.1 Preparation of B cells

B lymphocytes must first be isolated from peripheral blood which is the most convenient source of stimulated B cells. This is achieved by isolating the mononuclear cells using density centrifugation. If desired, T cells may then be removed initially using appropriate techniques, e.g. magnetic bead separation. Feeder cells will only be required if purified B cells are used. Where cells are to be obtained from spleen a 1 cm^2 piece of tissue is washed thoroughly with medium and then spleen cells teased out to a single cell suspension using two pairs of sterile forceps. Mononuclear cells are then prepared using lymphocyte separation medium as described for peripheral blood in *Protocol 10*.

Protocol 10. Isolation of mononuclear cells from blood

1. 20 ml of freshly collected anti-coagulated blood is diluted with an equal volume of Iscove's modified DME medium (IMDM) and dispensed into four, 20 ml sterile universals.
2. Using a syringe with a sterile cannula attached, carefully layer 8 ml of lymphocyte separation medium (Lymphoprep, Nycomed or Ficoll-Paque, Pharmacia) beneath 10 ml of the diluted blood.
3. Centrifuge at 400 *g* for 40 min at room temperature.
4. Using a Pasteur pipette, carefully draw off the mononuclear cell layer without removing an excessive volume of lymphocyte separation medium. Pool the mononuclear cell suspensions. Fill the tube with medium and centrifuge at 500 *g* for 10 min to wash the cells.
5. Remove supernatant and re-suspend cells. Dilute to 20 ml with medium and count the number of mononuclear cells. (Platelets may then be removed by centrifugation through 20% sucrose.)

7.2 Epstein–Barr virus (EBV)

Transforming EBV is harvested from cultures of the marmoset cell line B95-8 (41). Detailed procedures for the preparation of EBV supernatant from B95-8 cells, the testing of the supernatant in a cord blood assay, and an assay for EB nuclear antigen are given by Walls and Crawford (42) which is recommended reading. The supernatant should be stored at −70°C. In order to prevent the regression of cultures from EBV seropositive donors due to T cell cytotoxicity, it is necessary to include either the T-cell mitogen, phytohaemagglutinin (PHA), or cyclosporin in the medium if mononuclear cells containing T cells are transformed (43).

Protocol 11. EBV transformation

1. Dilute thawed viral supernatant with an equal volume of fresh medium containing 10% FCS.
2. Centrifuge mononuclear cell suspension at 500 *g* for 10 min.
3. Re-suspend the mononuclear cells in the virus-containing medium at 2×10^6 cells/ml. Incubate at 37°C in a 5% CO_2 atmosphere for 2 h with occasional gentle mixing.
4. Centrifuge at 500 *g* for 5 min. Remove supernatant and resuspend cells at 5×10^5 cells/ml in medium plus 20% FCS. Add 1% (v/v) PHA (Wellcome).
5. Dispense cell suspension into sterile tissue culture plates (either 2 ml or 0.2 ml wells are satisfactory) and incubate at 37°C in an atmosphere of 5% CO_2.

Protocol 11. *continued*

6. Cultures should be examined daily for evidence of outgrowth of clones of transformed cells. After 7 days (or earlier if medium turns yellow) replace half the medium with fresh medium plus 20% FCS. Spent medium may be assayed for antibody activity.
7. When the cells are almost confluent they can be expanded into larger volumes of fresh medium. The dilution factor should not exceed fourfold. Established lines may be conveniently maintained at densities between 2 and 20×10^5 cells/ml.

7.3 Maintenance of lymphoblastoid cell lines

Cells should be cultured when the medium turns yellow indicating a drop in pH. By this time clumps of cells will be visible. Generally, cells will require subculture every 3 to 5 days.

Protocol 12. Maintenance of lymphoblastoid cell lines

1. Examine the cells with the inverted microscope and note their appearance.
2. Transfer cell suspension to a centrifuge tube. Centrifuge at 400 *g* for 10 min.
3. Re-suspend cells in 10 ml of medium/FCS. Ensure that cells are very well mixed and immediately take a small aliquot (0.1 ml) to count viable cells in a haemocytometer (see Chapter 1).
4. Calculate the required volume of cell suspension to set cells up at between 2 and 5×10^5/ml. Add medium/FCS to 10 ml for a 25 cm^2 flask or 50 ml for a 75 cm^2 flask.
5. Spent supernatant may be assayed for antibody activity.

7.4 Cloning of lymphoblastoid cell lines

The major differences between cloning lymphoblastoid cell lines and hybrids are that the lymphoblastoid cells are more difficult to grow at a single cell density and, because of their marked tendency to clump in suspension, it is more difficult to ensure that a single cell suspension has been obtained. *Protocol 13* will result in six plates: one with 100 cells per well; one with 20 cells per well; two with 5 cells per well; and another two with 1 cell per well.

Protocol 13. Cloning of lymphoblastoid cell lines

1. Twenty-four hours before cloning, six flat-bottomed 96-well plates should be seeded with feeder cells as described in *Protocol 4*. Use only the inner 60

wells of the plate, filling the outer wells with medium or antibiotic solution. Either mouse peritoneal macrophage cells or human mononuclear cells, irradiated with 2500 rad, at 10^5 cells per well may be used as feeders (the former are preferred). Place plates in a 37°C incubator with 5% CO_2 atmosphere.

2. Cloning should be done from a culture in the mid-log phase of growth. Centrifuge the cell suspension of such a culture at 500 *g* for 5 min. Re-suspend the cells by brisk tapping and add 10 ml of medium plus 10% FCS. Mix the cell suspension very thoroughly and count cells immediately, also determining viability.
3. Calculate the dilution required to bring the cells to a concentration of 1000 cells/ml. Prepare a master dilution of at least 10 ml of a suspension containing 1000 cells/ml in medium plus 20% FCS.
4. Place 100 μl of the master dilution into the 60 wells of the first plate (100 cells/well).
5. Dilute five times by adding 2 ml of the master dilution to 8 ml of medium plus 20% FCS. Plate 100 μl into each well of the 20 cell well plate.
6. Dilute 20 times by adding 1 ml of the master dilution to 19 ml of medium plus 20% FCS. Plate 100 μl into each well of the 25 cell/well plates.
7. Finally dilute the master dilution 100 times and plate as above into the 1 cell/well plates.
8. Incubate the plates at 37°C under 5% CO_2.
9. Feed cultures after 7 days with medium plus 20% FCS and then as required according to growth. Expand clones by no more than fourfold dilution. When sufficient cells have been obtained freeze a number of vials of each line.

Note: with some lines it is difficult to obtain growth at a density of less than 20 cells per well.

7.5 Fusion procedures for human monoclonal antibodies

As stated a number of different fusion partners have been used by different researchers. The results have proved variable but we have found that fusion with the mouse myeloma cell, X653, produces the best results. Fusion conditions will need to be optimized for each laboratory and according to the myeloma line used for fusion.

Protocol 14. Fusion for human mAb production

1. Both cell lines should be in the mid-log phase of growth. Harvest the mouse myeloma cells (X653 is the preferred choice) and the lymphoblastoid

Protocol 14. *continued*

cells by centrifugation from culture medium. Count the cells and determine viability. Decide on the total number of cells to be fused. A fusion ratio of 1:1 should be used. Up to 30×10^6 cells of each type can be handled with ease.

2. Pour the required volume of each cell suspension into separate tubes. Make up to 25 ml with medium and centrifuge at 500 *g* for 7 min.
3. Repeat the washing step.
4. Re-suspend cells in 10 ml of medium, mixing thoroughly. Count the cells. For controls, plate four wells each containing myeloma cells and lymphoblastoid cells at 2×10^5 cells/ml in a 96-well plate.
5. Add the lymphoblastoid cells to the myeloma cells and centrifuge together at 400 *g*.
6. Remove the supernatant completely. Re-suspend the cells. Warm PEG to 37°C and add 1 ml slowly to the cells whilst agitating gently in a 37°C water bath. After 2 min exposure to PEG dilute slowly by adding 5 ml PBS (37°C) over 2 min and then 10 ml over the following minute.
7. Remove the supernatant from the cells and re-suspend the cells in 20% FCS medium at 5×10^5 cells/ml. Incubate in a 75 cm^2 flask overnight at 37°C, 5% CO_2.
8. Prepare a feeder layer of mouse peritoneal macrophage cells. Plate at 10^4 cells per well in medium plus 20% FCS containing 1 μM ouabain and HAT in 96-well flat-bottomed plates.
9. Next day add ouabain and HAT to the fusion cell suspension and plate 100 μl/well, i.e. a final cell concentration of 2.5×10^5/ml is achieved. Replace half the 2.5×10^5/ml supernatant of two control wells for each of the cell lines with medium containing HAT and ouabain at the standard concentrations. Incubate plates as above.
10. After 5 days feed the cultures by replacing half the medium with 20% FCS medium containing 1 μM ouabain and HAT. The cells in control wells without selective medium should have grown to confluence, whereas the cells fed with selective medium should appear to have died.
11. Once there are sufficient hybrids in the wells supernatant should be assayed for antibody activity. Where lymphoblastoid cells secreting high levels of antibody have been fused it will be necessary to ensure that this does not complicate the assay for specific hybrid antibody. Hybrids should be cloned as soon as practical. Not all of the mouse–human hybrids are likely to be stable producers of antibody.

    Aminopterin and ouabain may be eliminated from the medium after 3 weeks.

8. Antibody production

Kilogram quantities of mAbs are now produced each year for many different applications. This quantity is manufactured by many different companies and laboratories by a variety of different techniques. The commercial production of mAbs is dealt with in Chapter 10 and this section is mainly designed to instruct the reader how to produce small quantities of mAbs. The first choice to be made by the producer is whether to use *in vivo* or *in vitro* methods.

8.1 *In vivo* production

Small volumes containing relatively high concentrations of murine antibodies, i.e. 1–10 mg/ml, can be obtained by injecting hybridoma cells secreting the desired antibody into the peritoneal cavity of mice where they grow, generating ascites fluid. In some cases the procedure is not successful and then adjuvants such as trypan blue, pristane, or incomplete Freund's adjuvant should be used. Balb/C mice should be used for production of mouse mAbs because the parent myelomas are of Balb/C origin and would not be compatible with other strains.

Protocol 15. Production of ascitic fluid

1. If adjuvants are required mice should be pre-treated with either
   (a) 0.5 ml pristane (2, 6, 10, 14-tetramethylpentadecane) injected i.p. 10–14 days before injection of the hybridomas, or
   (b) 0.4 ml trypan blue injected i.p. 24 h before cell inoculation and then inject 0.2 ml trypan blue i.p. 1 h before injecting the hybridoma cells. The trypan blue is prepared by dissolving 2 g in a minimal amount of distilled water and dialysing against distilled water for 24–48 h with two changes of water per day. The volume is then adjusted to 200 ml (10 mg/ml) and 1.7 g NaCl added (8.5 mg/ml).
2. Inject 5×10^6 viable hybridoma cells per mouse i.p. contained in 0.2 ml culture medium.
3. Monitor mice daily. After 7–14 days the abdomen should become swollen. We usually aim to collect 5 ml ascitic fluid per mouse.
4. Asphyxiate the mouse with CO_2, swab the abdomen, and expose the peritoneal wall (see *Protocol 4*).
5. Remove up to 5 ml of ascitic fluid using a 21 g needle attached to a syringe.
6. The ascitic fluid may then be purified or aliquoted and frozen at −70°C

Protocol 15. *continued*

until required for use. An aliquot can be used to re-inject other mice but it is important to monitor antibody titres in case the population is overgrown by non-secretors.

8.2 *In vitro* production

Hybridoma cells grow freely in suspension culture and thus scale-up from 2 ml to 10 ml to 100 ml to 1 litre culture and beyond is relatively easy. Two basic strategies can be adopted for the *in vitro* production of mAbs and they are: conventional cell cultures involving large volumes and low cell densities, for example using air-lift or stirred tank bioreactors; and culture systems employing relatively small volumes and high cell densities, for example perfusion cultures using hollow-fibres (see Chapter 7). In addition, a variety of other processes have been developed for commercial large-scale production during recent years and these are discussed in Chapter 10. Several hollow-fibre systems are available commercially and for a laboratory without fermentation experience this may be an easier alternative than cell culture scale-up. In normal flask cultures hybridoma cells produce 5–50 μg antibody/ml. Thus antibody yields from conventional *in vitro* culture are significantly less than those obtained from the *in vivo* technique. For example, a 5 ml sample of ascites should generate approximately 50 mg of antibody whereas it would probably require a 5 litre suspension culture to generate 50 mg, assuming the culture contains 10 μg antibody/ml.

The larger volumes associated with *in vitro* culture are a further disadvantage when the antibody has to be purified, although this problem is fairly easily dealt with using ultrafiltration concentration devices to concentrate the culture medium prior to purification. For the production of human hybridomas, *in vitro* culture is a necessity. Due to the problems of adventitious viruses associated with rodents, it is also preferable to produce murine mAbs intended for therapeutic use by *in vitro* technology. The production process should try to ensure batch-to-batch consistency by starting each production run of the hybridoma from a frozen vial of cells. Thus a large bank of vials should be established in liquid nitrogen as soon as possible after the third cloning of the original hybrid. The frozen vial is thawed and a 10 ml flask culture initiated. This culture is grown and inoculated directly into a 100 ml culture in a spinner flask or expanded into a 50 ml flask culture which can then be inoculated into a 500 ml spinner culture. This culture can then be used to inoculate a 2 or 5 litre culture. At each step in the scale-up procedure it is important to maximize cell viability by keeping cell numbers at about 5–8×10^5/ml, thus maintaining cells in the mid-log phase of growth. In the final production vessels the cells can be grown to extinction although it is advisable to establish the basic kinetics of antibody production versus cell growth to optimize the time of harvest. In addition, antibody yield may be improved in such batch culture systems by extending cell viability using careful monitoring and control of the physicochemical environment. Extra nutrients can be pulse-fed to the culture to prolong cell viability. We have successfully grown hybridoma cells in spinner flasks, stirred tank reactors, and air lift reactors (see Chapter 7). Antibiotics should be avoided if possible and β-lactam agents should not be used in the production of antibodies intended for therapeutic use.

9. Future prospects

Ten years ago it would have been difficult to predict the enormous impact that mAbs have had on science and medicine and on the rapid growth of the new biotechnology industries. Future predictions are therefore fraught with dangers. During the past few years molecular biological techniques have been increasingly applied to mAb technology. In this fast-moving field it is difficult to speculate but already it has proved possible to modify the antibody molecule extensively, for example:

(a) hybrid antibodies which have different binding specificities so that they can form bridges between two different chemical structures;

(b) the production by molecular engineering of simple chimeric antibodies with mouse variable regions and human constant regions (44, 45); these antibodies are preferred for therapy since they should prove less immunogenic in man and less likely therefore to produce side-effects;

(c) the formation of antibodies which retain only those amino acid residues from the mouse antibody essential for the antigen-binding site, while the remainder of the antibody is human (46, 47). This approach allows for the grafting of different antigen-binding sites on to a human antibody stalk containing the Fc receptor with effector functions appropriate for the intended use of the antibody. The Fc region of the antibody is situated on one of the three fragments generated following protease (papain) treatment of the IgG molecule. The Fc (fragment crystallizable) lacks the ability to bind antigen but has important effector functions, e.g. there are specific Fc receptors on the macrophages which bind IgG molecules coating bacteria thus promoting phagocytosis.

Finally, the recent finding (48) that single-chain antibodies produced in bacteria have antigen specificity is an exciting development which may lead to the use of bacteria as rival production vehicles to mammalian cells. At this stage this would appear to be some time down the track and, for the foreseeable future, hybridoma cells will provide the means of producing large quantities of antibodies.

Recommended reading

Engelman, E. G., Foung, S. K. H., Larrick, J., and Raubitschek, A. (ed.) (1985). *Human Hybridomas and Monoclonal Antibodies*. Plenum Press, New York.
Goding J. W. (1986). *Monoclonal Antibodies: Principles and Practice*. Academic Press, London.
Newell, D. G., McBride, B. W., and Clark, S. A. (1988). *Making Monoclonals*. Public Health Laboratory Service, The Laverham Press, Salisbury.
Zola, H. (1987). *Monoclonal Antibodies: A Manual of Techniques*. CRC Press, Boca Raton, Florida.
Langone, J. L. and Van Vanakis, H. (ed.) (1986). *Methods in Enzymology*, Vol. 121. Academic Press, New York.

References

1. Burnet, F. M. (1957). *Austral. J. Sci.* **20,** 67.
2. Cotton, R. G. H. and Milstein, C. (1973). *Nature* **244,** 42.
3. Schwaber, J. and Cohen, E. P. (1974). *Proc. natl Acad. Sci. USA* **71,** 2203.
4. Köhler, G. and Milstein, C. (1975). *Nature* **256,** 495.
5. James, K. and Bell, G. T. (1987). *J. immunol. Methods* **100,** 5.
6. Olsson, L. and Kaplan, H. S. (1980). *Proc. natl Acad. Sci. USA* **77,** 5429.
7. Olsson, L., Kronstrom, H., Cambon-De Monzon, A., Honsik, C., Brodin, T., and Jakobsen, B. (1983). *J. immunol. Methods* **61,** 17.
8. Croce, C. M., Linnenbach, A., Hall, W., Steplewski, Z., and Koprowski, H. (1980). *Nature* **288,** 488.
9. Kozbor, D., Legarde, A. E., and Roder, J. C. (1982). *Proc. natl Acad. Sci. USA* **79,** 6651.
10. Edwards, P. A. W., Smith, C. M., Neville, A. M., and O'Hare, M. J. (1982). *Eur. J. Immunol.* **12,** 641.
11. Kozbor, D., Tripputi, P., Roder, J. C., and Croce, C. M. (1984). *J. Immunol.* **133,** 3001.
12. Teng, N. N. H., Lam, K. S., Riera, F. C., and Kaplan, H. S. (1983). *Proc. natl Acad. Sci. USA* **80,** 7308.
13. Kearney, J. F., Radbruch, A., Liesegany, B., and Rajewsky, K. (1979). *J. Immunol.* **123,** 1548.
14. Galfrè, G. and Milstein, C. (1981). In *Methods in Enzymology*, Vol. 73 (ed. J. J. Langone and H. Van Vanakis), p. 3. Academic Press, New York.
15. Shulman, M., Wilde, C. D., and Köhler, G. (1978). *Nature* **276,** 269.
16. Köhler, G., Howe, S. C., and Milstein, C. (1976). *Eur. J. Immunol.* **6,** 292.
17. Galfrè, G., Milstein, C., and Wright, B. (1979). *Nature* **277,** 131.
18. Kilmartin, J. V., Wright, B., and Milstein, C. (1982). *J. Cell. Biol.* **93,** 576.
19. Littlefield, J. W. (1964). *Science* **145,** 709.
20. Zimmerman, U., Schmitt, J. J., and Kleinhaus, P. (1988). In *Clinical Applications of Monoclonal Antibodies* (ed. R. Hubbard and V. Marks), p. 3. Plenum Press, New York.
21. Oi, V. T. and Herzenberg, L. A. (1980). In *Selected Methods in Cellular Immunology* (ed. B. B. Mishell and S. M. Shiigi), p. 351. W. H. Freeman, San Francisco.

22. Zola, H. and Brooks, D. (1982). In *Monoclonal Hybridoma Antibodies: Techniques and Applications* (ed. J. G. Hurrell), p. 1. CRC Press, Boca Raton, Florida.
23. Goossens, D., Champomier, F., Rouger, P., and Salmon, C. (1987). In *Monoclonal Antibodies against Human Red Blood Cells and Related Antigens* (ed. P. Rouger and C. Salmon), p. 23. Libraire Arnette, Paris.
24. Schwaber, J. F., Posner, S. F., Schlossman, S. F., and Lazarus, H. (1984). *Human Immunol.* **9,** 137.
25. Kozbor, D., Roder, J. C., Sierzega, M. E., Cole, S. P. C., and Croce, C. M. (1986). In *Methods in Enzymology*, Vol. 121 (ed. J. L. Langone and H. Van Vanakis), p. 120. Academic Press, New York.
26. Edwards, P. A. W. and O'Hare, M. J. (1986). In *Monoclonal Antibodies* (ed. P. C. L. Beverley), p. 60. Churchill Livingstone, Oxford.
27. Melamed, M. D., Gordon, J., Ley, S. J., Edgar, D., and Hughes-Jones, N. C. (1985). *Eur. J. Immunol.* **15,** 742.
28. Crawford, D. H. (1985). In *Human Hybridomas and Monoclonal Antibodies* (ed. E. G. Engelman, S. K. H. Foung, J. Larrick, and A. Raubitschek), p. 37. Plenum Press, New York.
29. Goossens, D., Champomier, F., Rouger, P., and Salmon, C. (1987). *J. immunol. Methods* **101,** 193.
30. Winger, L., Winger, C., Shastry, P., Russell, A., and Longenecker, M. (1983). *Proc. natl Acad. Sci. USA* **80,** 4484.
31. Doyle, A., Jones, T. J., Bidwell, J. L., and Bradley, B. A. (1985). *Human Immunol.* **13,** 199.
32. Bradshaw, P. A., Perkins, S., Lennette, E. T., Rowe, J., and Foung, S. K. H. (1988). In *Clinical Applications of Monoclonal Antibodies* (ed. R. Hubbard and V. Marks), p. 149. Plenum Press, New York.
33. Zimmermann, U., Schultz, J., and Pilwat, G. (1973). *Biophys. J.* **13,** 1005.
34. Spitz, M., Spitz, L., Thorpe, R., and Eugui, E. (1984). *J. immunol. Methods* **70,** 39.
35. Stahli, C., Staehelin, T., Miggiano, V., Schmidt, J., and Haring, P. (1980). *J. immunol. Methods* **32,** 297.
36. Möller, S. A. and Borrebaeck, C. A. K. (1988) In *In Vitro Immunization in Hybridoma Technology* (ed. C. A. K. Borrebaeck), p. 3. Elsevier, Amsterdam.
37. Borrebaeck, C. A. K. (ed.) (1988). In *In Vitro Immunization in Hybridoma Technology*, p. 209. Elsevier, Amsterdam.
38. Gilbert, K. and Dresser, D. W. (1987). In *Lymphocytes: A Practical Approach* (ed. G. G. B. Klaus), p. 109. IRL Press, Oxford.
39. Newell, D. G., McBride, B. W., and Clark, S. A. (ed.) (1988). *Making Monoclonals.* Public Health Laboratory Service, The Laverham Press, Salisbury.
40. Engvall, E. and Perlmann, P. (1971). *Immunochemistry* **8,** 871.
41. Miller, G. and Lipman, M. (1973). *Proc. natl Acad. Sci. USA* **70,** 190.
42. Walls, E. V. and Crawford, D. H. (1987). In *Lymphocytes: A Practical Approach* (ed. G. G. B. Klaus), p. 149. IRL Press, Oxford.
43. Neitzel, H. (1986). *Human Genet.* **73,** 320.
44. Neuberger, M. S., Williams, G. T., Mitchell, E. B., Jomhal, S. S., Flanagan, J. G., and Rabbits, T. H. (1985). *Nature* **314,** 268.
45. Morrison, S. L., Johnson, M. J., Herzenberg, S. A., and Oi, V. T. (1984). *Proc. natl Acad. Sci. USA* **81,** 6851.

46. Jones, P. T., Dear, P. H., Foote, J., Neuberger, M. S., and Winter, G. (1986). *Nature* **321,** 522.
47. Reichmann, L., Clark, M., Waldmann, H., and Winter, G. (1988). *Nature* **332,** 323.
48. Ward, E. S., Güssow, D., Griffiths, D. A. D., Jones, P. T., and Winter, G. (1989). *Nature* **341,** 544.

7

Bioreactors for mammalian cells

ANITA HANDA-CORRIGAN

1. Introduction

A plethora of bioreactors for the cultivation of mammalian cells are currently available on the market (1–4). They range from modified bacterial vessels of 1–10 000 litres (e.g. stirred tank and airlift bioreactors) to compact, process-intensified systems such as hollow fibre and flat-membrane bioreactors. The choice of bioreactor for a particular cell is determined by a number of factors: the cell type; the nature of the product; the scale of operation; availability of space and services; and, the capital and operational costs of the equipment.

In general, cell culture bioreactors can be categorized into two types—those used for the cultivation of anchorage-dependent cells (e.g. primary cultures derived from normal tissues and diploid cell lines) and those used for the cultivation of suspended mammalian cells (e.g. cell lines derived from cancerous tissues and tumours, transformed diploid cell lines, hybridomas, etc.). In some cases the same bioreactor may be adapted or modified to grow both anchorage-dependent and suspended cells. Ideally, any cell culture bioreactor must be able to maintain a sterile culture of cells in medium conditions which maximize cell growth and/or productivity. In practice this has often been difficult to achieve for the following reasons.

(a) Contamination problems can occur at all stages of operation. At the early stages of expansion of a culture for inoculum preparation, operator errors and excessive dependence on anti-microbial agents contribute to increased risks of contamination. During larger-scale cultivation, problems with maintaining sterility can occur in a number of areas in the bioreactor including: exit and entry ports; air inlet and exhaust filters; valves and seals; bioreactor internals such as agitator shafts and probes; connectors and connecting lines to other vessels, pumps, analysers; etc. Successful sterile cultivation at any scale requires a combination of highly skilled cell culturists and well-designed (but not over-designed!) bioreactors. Care should be taken not to make a choice based solely on the bioreactor's capability of achieving high cell or product yields. Some of the more sophisticated mammalian cell culture bioreactors are often prone to high contamination risks.

(b) A number of complex interactions (physical, chemical, and biochemical) coexist in a bioreactor. These include mixing and aeration; oxygen demand and supply; cell growth and product formation kinetics in relation to the nutritional requirements of the cells; waste and toxic factor accumulation; and detrimental effects to the culture due to the mixing or aerating devices used. All these factors have to be considered in relation to the scale, geometry, and configuration of the bioreactor. In practice, even careful consideration of all these variables often results in optimum rather than maximum yields of products or cells.

At present there are some 30 or more slightly modified or novel designs of bioreactors for mammalian cells. In the following sections the theoretical and practical aspects of cultivating cells in some succesful bioreactor designs will be described.

2. Basic bioreactor configurations: batch systems

In batch cultivation, an inoculum of known density is seeded into a specified volume of pre-conditioned medium in the bioreactor. Ideally, nothing more should be added or removed from the bioreactor during the course of the cultivation. However, in practice, additions of air (for oxygen supply) and CO_2 and acids or bases (for pH control) are made at scales of 1 litre and above. Batch cultivation of suspension cells can be carried out in two types of bioreactors: stirred tank and airlift reactors. Their basic principles of operation and design features (5, 6) are described in *Figures 1* and *2*, and in *Tables 1* and *2*, respectively.

Protocol 1. Batch cultivation of suspended mammalian cells in a bench-top stirred tank reactor

Equipment

- Cell culture stirred tank reactor (e.g. 1 litre working volume)
- Measurement and control instrumentation for pH, dissolved oxygen tension (DOT), temperature, and agitation speed (r.p.m.)
- Autoclave (preferably a high-vacuum type)
- pH probe
- Oxygen probe
- Temperature probe
- Hydrophobic air filters (0.2 μm pore size)
- Sampling bottles (5 ml bijoux)
- Silicone tubing
- Air cylinder

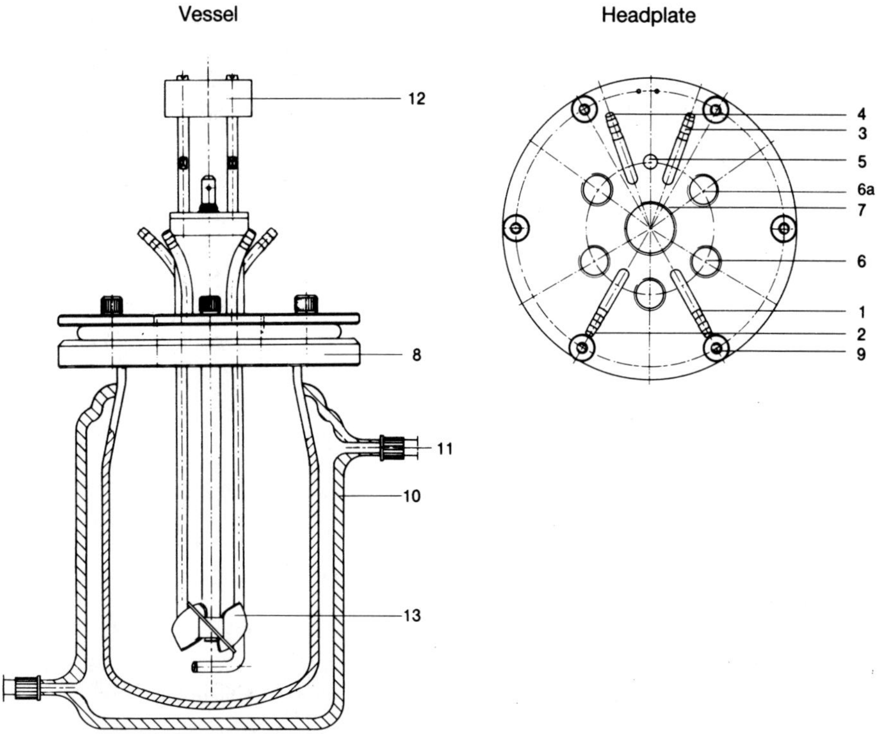

Figure 1. A 2 litre stirred tank bioreactor for mammalian cells. 1, Sample pipe; 2, air inlet pipe with porous sparger; 3, medium inlet pipe; 4, air outlet; 5, thermometer pocket; 6, four sensor/auxiliary ports; 6a, threaded auxiliary port; 7, threaded port for stirrer assembly; 8, clamping ring; 9, six mill nuts; 10, jacketed 2 litre vessel; 11, hose connector; 12, stirrer assembly; 13, impeller. (Reproduced with kind permission from FT Applikon Ltd, UK.)

- CO_2 cylinder (or 5% CO_2 in air supply)
- N_2 cylinder (oxygen-free nitrogen)
- Laminar air flow hood
- Haemocytometer
- Microscope

Materials

- 1 litre sterile, basal cell culture medium (e.g. RPMI 1640)
- 50 ml new-born calf serum or fetal calf serum (FCS)
- 100 ml concentrated, suspended mammalian cells (e.g. BHK or hybridoma cells) at a density of 1×10^6 cells/ml (i.e. total number of cells = 1×10^8). Cell viability >95% must be used

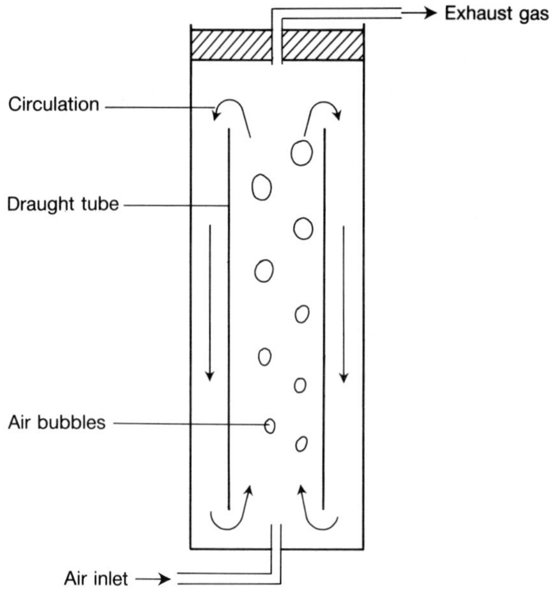

Figure 2. Air lift bioreactor for the cultivation of suspended mammalian cells.

Table 1. Principal features of stirred tank bioreactors for mammalian cells

Function: To achieve a homogeneous mixture of the bioreactor contents with a mechanical agitator (the impeller).

Mixing: For vessels of up to 500 litres, low shear flat bladed or marine propellers are used.
For larger vessels, high shear turbine type impellers are used.

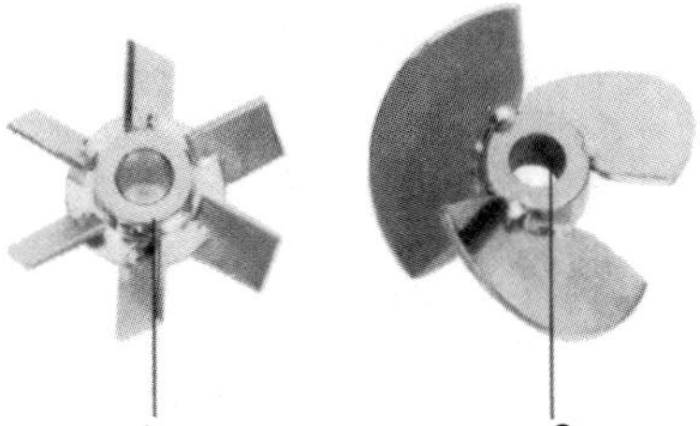

1. 6-bladed flat disc turbine
2. 3-bladed marine impeller

1. 2.

reproduced with kind permission from F. T. Applikon, Ltd, UK.

Oxygenation: Direct sparging of air bubbles into the bioreactor via sintered discs, open pipes, or porous rings. Often problems with foaming and cell damage can be caused due to sparging. Alternative bubble-free aerators may be used (e.g. silicone tubing and caged aerators).

Sterilization: 1–10 litres vessels can be autoclaved for 20–30 min at 121 °C, 15 psi (lb/in^2). Larger vessels have to be sterilized *in situ* with steam at 121 °C and 15 psi steam pressure.

Protocol 1. *continued*

- pH buffer solutions 4, 7, and 9
- Non-toxic detergent (e.g. 7×, Flow Laboratories)
- Cell-culture quality water

Methods

Cleaning and sterilization of the bioreactor

1. Carefully remove probes and headplate from the vessel.
2. Empty out vessel contents and clean the inside thoroughly with 7× detergent. Rinse thoroughly with tap water and finally with cell-culture quality water.
3. Repeat cleaning procedure for headplate, impeller, sparge ring, and inlet and exit ports.
4. Check all 'O' rings, gaskets, and seals. Replace old or damaged ones.
5. Using silicone tubing, connect new air-filters to ports in the headplate for headspace gassing of air, for sparging of air, N_2, and CO_2 into the bioreactor, and for exhaust gases.

Table 2. Airlift bioreactors for hybridoma cultivation

Principle of operation. Gas is introduced at the bottom of the vessel, within the draught tube. A reduction in the density of the aerated contents in the draught tube results in a circulation of the culture through the draght tube, and down in the outer zone of the vessel.

Advantages: no moving parts or mechanical seals; adequate oxygen transfer (0.2–1 mmol/litre/h); low hydrodynamic shear forces; low power input per unit volume (10–15 watt/m^3).

Operation: batch operation. Temperature controlled at 37 °C via a circulating water jacket. pH controlled by automatic addition of CO_2 into the sparged gas, or by sodium hydroxide addition. Dissolved oxygen tension controlled by varying the concentration of oxygen in air. Foaming may be controlled by the addition of antifoam agentin conjunction with Pluronic F–68.

Scale-up. 10–2000 litres airlift bioreactors are currently being used for hybridoma cultivation (Celltech Ltd). Improved oxygen transfer rates have been shown at the larger scales. The increase in hydrostatic pressure resulting from increased bioreactor height does not have deleterious effects on the cells.

Media. Media supplemented with 2–10% animal sera as well as defined serum-free media have been used successfully.

Hybridoma growth. Hybridoma cell lines of mouse, rat, and human origin have been cultivated batch-wise for 10–17 days, depending on the particular cell line. Hybridoma growth kinetics are not affected by dissolved oxygen tension in the range of 10–100% saturation.

Problems. A major problem at the small scales is the accumulation of foam in which cells and product are lost.

Protocol 1. *continued*

6. Check that all measuring probes are functional. Calibrate the pH probe against pH buffer solutions 4, 7, and 9.
7. Put the probes and headplate back on to the bioreactor and tighten all nuts and clamping rings. Ensure that all unused ports are blanked-off.
8. Autoclave for 20 min at 121 °C, 15 psi.
9. Remove from the autoclave when temperature is below 30–40 °C so that the bioreactor can be safely carried out.
10. Place the cooled reactor in a laminar air flow hood.

Medium addition and calibration of the dissolved oxygen probe

1. Using aseptic techniques, pour 850 ml sterile basal medium through a large addition port on the bioreactor headplate.
2. Seal off the medium addition port and transfer bioreactor out of the laminar air flow hood to a bench where measurement and control modules have been set up.
3. Connect up all probe and impeller leads to the appropriate modules.
4. Set the impeller speed at 50 r.p.m. and the temperature at 37 °C.
5. Calibrate the dissolved oxygen probe; sparge the basal medium for approximately 20 min with oxygen-free N_2. Set-point the oxygen meter to zero. Next, sparge the basal medium with air for another 20 min. Set-point the oxygen calibration meter to 100%. Repeat the above sequence of steps once more to ensure calibration is accurate.

Inoculum and serum addition

1. Transport bioreactor back into the laminar air flow hood.
2. Add 100 ml concentrated suspended cells (at 1×10^6 cells/ml; >95% viability) and 50 ml serum, into the bioreactor, via an inoculum addition port.
3. Seal the inoculum addition port, take the bioreactor out of the hood, and set-up on the bench.
4. Set-point cultivation parameters on the bioreactor control modules: pH at 7.0–7.1; temperature at 37 °C; DOT at 10–40%; Stirrer speed at 50 r.p.m.

2.1 Temperature and pH control

For temperature control in bioreactors with a circulating water jacket (e.g. LH Fermentation) connect the jacket to a thermocirculating water bath. For bioreactors with circulating warm air jacket (e.g. SGi, France) switch on the hot air supply.

Cell culture medium (e.g. RPMI 1640) containing sodium bicarbonate buffer can be used to maintain pH in range of 7.0–7.2 by the addition of CO_2

$$H_2O + CO_2 = H_2CO_3 = H^+ + HCO_3^-. \quad (1)$$

At an initial inoculum density of about 1×10^5 cells/ml, sparging of air (for oxygenation purposes) also results in flushing of CO_2 out of the culture medium and thus the pH increases. This can be counteracted by a low sparge of CO_2 in the medium. As the cells begin to grow, lactate is produced which results in a decrease in pH and, therefore, the requirement for CO_2 decreases. With further growth, the culture pH may drop even further and the requirement for CO_2 ceases completely. This drop in pH may be counteracted by the addition of 1 M NaOH. Take care that the addition of base is slow so that large localized pH increases (which can result in cell death) in the culture do not occur. During the death phase of the culture, the pH begins to rise gradually.

In order to control pH by CO_2 addition:

(a) Connect the CO_2 supply to a solenoid valve in the control module and a CO_2 sparge line from the control module to the bioreactor. CO_2 will be supplied on demand.

(b) Connect an additional line supplying air continuously to the headspace of the bioreactor. This will remove any excess CO_2 that accumulates in the headspace.

2.2 Dissolved oxygen control

The oxygen demand for most cells has been shown to fall in the range $2–10 \times 10^{-12}$ g O_2 per cell per h (or 0.06–0.3 μmol oxygen/10^6 cells/h). In a 1 litre bioreactor, minimal sparging of air may be required to meet the oxygen demand of the cells. To maintain the dissolved oxygen tension between 20 and 40%, air can be sparged into the bioreactor intermittently. In some cases, however, the sparging may result in excessive foam formation. When this occurs, alternative bubble-free aeration devices or foam stabilizing agents (such as Pluronic F-68) may have to be used.

(a) Connect the air supply to the solenoid valve in the control module.

(b) Connect the exit air supply from the solenoid valve to a rotameter and then to the air inlet filter of the sparge line to the bioreactor.

Sparged air will enter the culture intermittently to maintain the appropriate set-point.

2.3 Monitoring the cultivation

Cell growth can be monitored by aseptic sampling of the culture at regular intervals (every 24 h) for approximately 7–10; days. Samples can be analysed for cell numbers and viability, product concentration, nutrient (e.g. glucose and glutamine) consumption, and metabolic waste accumulation (e.g. ammonia and lactate). These data can be used to develop cell growth and product formation kinetics. A typical growth curve for a suspended mammalian cell cultivated in a laboratory-scale stirred tank reactor is shown in *Figure 3*. Repeated cultivation tests at the laboratory scale can be carried out to determine optimum growth and/or production conditions for a particular cell line. Generally, a total of 10^9 cells can be obtained from a single batch cultivation of suspended cells in a 1 litre bioreactor.

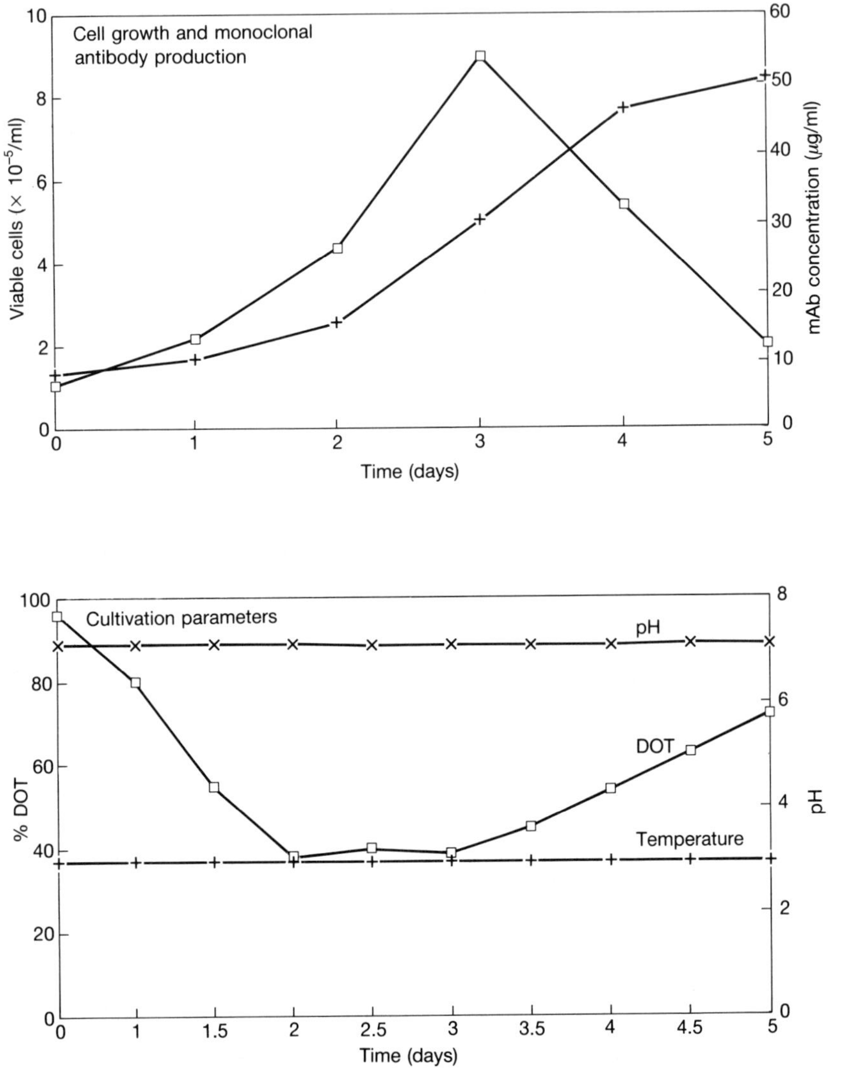

Figure 3. Batch cultivation of hybridoma cells in a 2 litre stirred tank reactor (SGi, France).

2.4 Scaling-up

The scaling-up of stirred tank bioreactors requires that the suspended cells, nutrients, and gases are maintained homogeneously mixed and that the oxygen demand of the cells is adequately met by the sparged air supply. In practice, scale-up is not always easy to achieve due to foaming and cell damage problems associated with sparging and mixing. It is recommended that the mixing and aerating properties of the bioreactor are considered independently.

2.4.1 Mixing

An ideal batch reactor is perfectly mixed with uniform circulation patterns (7). The impellers used for mixing cell cultures are either axial flow impellers (e.g. marine propellers) or radial flow impellers (e.g. turbine impellers). Axial flow impeller blades make less than a 90° angle with the plane of rotation, whilst the radial flow impeller blades are parallel to the axis of the drive shaft. When a low-viscosity culture (such as mammalian cell culture) is stirred by either of these types of impellers, a swirling flow pattern of the culture is obtained. At high mixing speeds, a vortex is produced due to centrifugal forces acting on the rotating liquid. As the vortex reaches the impeller, severe air entrainment and large fluctuating forces on the impeller shaft may occur. Laboratory-scale cell culture bioreactors are therefore mixed at 50–300 r.p.m., above which problems due to vortexing and air entrainment may become apparent. To achieve top to bottom circulation without vortexing, large-scale vessels are often fitted with baffles (*Figure 4*).

The presence or absence of turbulence in an impeller stirred vessel can be determined by the impeller Reynolds number, *Re*,

$$Re = \frac{\rho N D^2}{\mu} \qquad (2)$$

where ρ is the density of the fluid, N the agitator speed, D the impeller diameter, and μ the fluid viscosity. For $Re < 10$, the flow is laminar and for

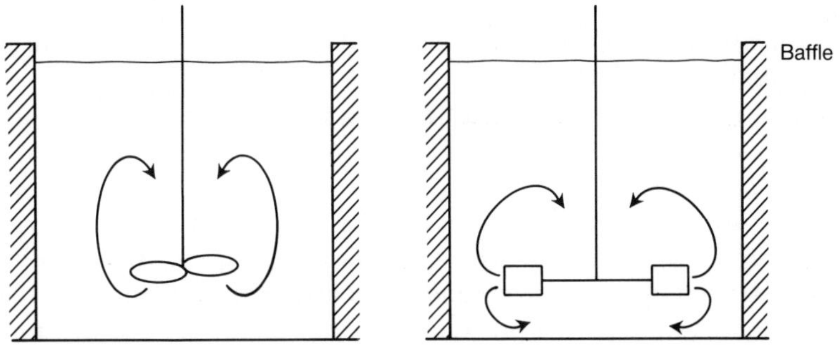

Figure 4. Liquid circulation patterns in baffled stirred tank reactors.

$Re > 2 \times 10^4$, the flow is turbulent. In between these values is a transition range where the flow is turbulent near the impeller and laminar in remote parts of the vessel. In the laminar flow regime, the power (P) imparted to the fluid is proportional to the fluid viscosity and independent of the fluid density

$$P \propto \mu N^2 D^3 \qquad \text{or} \qquad P_o \propto \frac{1}{Re}$$

Thus,

$$P = P_o\, \mu N^2 D^3 \tag{3}$$

where P_0 is a proportionality constant called the power number. In the turbulent flow regime, P is independent of fluid viscosity, but strongly dependent on impeller diameter

$$P \propto \rho N^3 D^5.$$

Thus,

$$P = P_0 \rho N^3 D^5 \tag{4}$$

Scale-up based on a fixed power input per unit volume (i.e. P/V) can be used for geometrically similar vessels,

$$\frac{P_S}{V_S} = \frac{P_L}{V_L} \tag{5}$$

where subscripts S and L refer to the small and large vessels, respectively. For example, under turbulent conditions,

$$N_S^3\, D_S^2 = N_L^3\, D_L^2. \tag{6}$$

The impeller speed for a large-scale bioreactor may therefore be calculated from the values of N_S, D_S, and D_L. Although equation 6 offers a simple method for scaling-up, it is only accurate for geometrically similar, ungassed vessels. In addition, one has to be aware that in using this method for scale-up other properties (e.g. circulation times and shear rates) in the small- and large-scale vessels may differ.

2.4.2 Aeration: scale-up on the basis of a constant volumetric mass transfer coefficient, K_La

The oxygen supply for mammalian cells can be achieved by various means.

Headspace aeration

In 0.1 to 5 litre bioreactors, oxygen may be supplied by simple diffusion of oxygen from air in the headspace. At a constant temperature and pressure, the concentration of oxygen in the medium in equilibrium with the headspace is proportional to the partial pressure of oxygen in the gas phase. Thus,

$$p = H c \tag{7}$$

where p is the partial pressure of oxygen in the gas phase, c the concentration of oxygen in the medium, and H is Henry's constant.

The concentration of oxygen in air-saturated medium is approximately 6.436 mg/litre (or 6.4 p.p.m.) at 37°C and normal atmospheric pressure. The equilibrium concentration of oxygen in the medium may be increased by increasing the partial pressure of oxygen in the gas phase. Care must be taken that elevated partial pressures do not result in oxygen toxicity problems for the cells.

Increasing the rate of oxygen transfer by sparging and/or agitation

The volumetric oxygen transfer rate (N_A) in a bioreactor is commonly expressed as

$$N_A = K_L a\,(C^* - C_L) \tag{8}$$

where N_A is the volumetric rate of oxygen transfer from the gas phase (say, inside a bubble), to the liquid phase (the medium), via a gas–liquid interface. The term $(C^* - C_L)$ represents a concentration difference or a driving force between the bulk of the liquid (C_L) and the interface (C^*). K_L is the rate coefficient and a is the interfacial area of the bubbles. The combined term $K_L a$ is also known as the volumetric mass transfer coefficient. An increase in the oxygen transfer rate can therefore be brought about by increasing K_L, a (or $K_L a$), or $(C^* - C_L)$.

In conventional sparged, stirred tank reactors, it is common practice to increase the oxygen transfer rate by increasing bubble break-up (and increasing the interfacial area) with the impeller. For mammalian cell culture systems, this increased bubble break-up results in excessive foam formation and losses of cells and proteins at the medium surface (8). Many cell culture systems have therefore been scaled-up by maintaining a constant $K_L a$ using bubble-free aeration devices. For example, in the ChemCell system (Chemap AG, Switzerland), pure oxygen or air is introduced into a vibrating gassing cylinder, enclosed by a stainless steel mesh (*Figure 5a*). The oxygen transfer rate is a function of the amplitude of vibration, the dissolved oxygen concentration, and the gas flow rate (*Figure 5b*). Bulk mixing in the stirred tank reactor is carried out with a bottom-driven impeller.

For mammalian cell culture, it is recommended that the impeller is used solely for mixing, and that a suitable aeration device is used to satisfy the oxygen demand of the cells (9).

2.5 Supports for anchorage-dependent cells

A number of substrates are available for cell attachment, spreading, and growth. The simplest batch cultivation systems are stationary flasks or rotating bottles made of plastic (e.g. polystyrene, polycarbonate, etc.). The plastic has to be wettable and surface treated to carry negative charges. Batch cultivation in packed bed reactors can also be carried out using high-surface

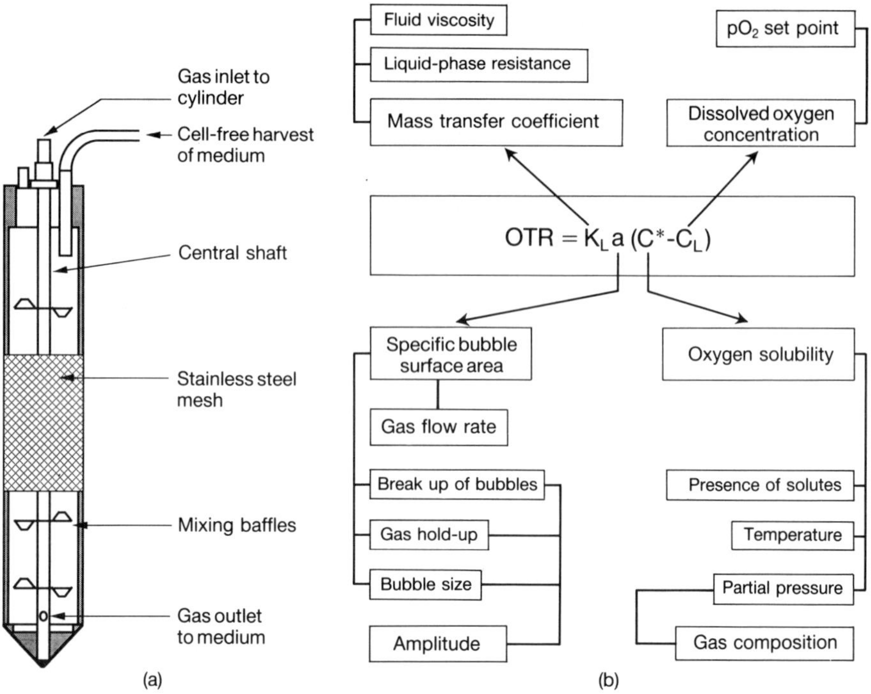

Figure 5. (a) Bubble-free aerator used in the ChemCell Bioreactor; (b) factors affecting the oxygen transfer rate (OTR) in the ChemCell gassing cylinder. (Reproduced with kind permission from Chemap AG, Switzerland.)

area packing materials such as sponges, steel springs, porous ceramic particles, calcium alginate gels, etc. Alternatively, cells may be grown on microcarriers (see Chapter 1) which are kept in suspension in stirred tank reactors (Section 3.2). *Protocol 2* describes the use of glass beads (3 mm diameter) for the growth of anchorage-dependent cells in a packed-bed reactor.

Protocol 2. Batch cultivation of anchorage-dependent cells in a 1 litre glass sphere propagator (with kind permission from Dr J. P. W. Whiteside)

Equipment

- 1 litre glass sphere propagator consisting of a 200 ml glass sphere resrvoir and 800 ml medium reservoir (*Figure 6*)
- 37°C incubator or hot room
- 5% CO_2 in air gas cylinder
- 1 litre sterile Duran bottles

- Sterile 1 litre glass beaker
- 2 sterile 100 ml centrifuge tubes

Materials

- 1 litre sterile, RPMI 1640 supplemented with 5–10 fetal or new-born calf serum and 6 p.p.m. silicone antifoam C (Sigma)
- 1×10^8 viable, anchorage-dependent cells (e.g. BHK monolayer cells)
- Cell-culture quality water
- 180 ml of 0.2% EDTA (Sigma) made up in PBS and 20 ml of 2.5% trypsin (Flow Laboratories).

Methods

Cleaning and sterilizing the bioreactor

1. Clean all glassware, beads, and silicone tubing with hot water and rinse with cell-culture quality water.
2. Assemble the bioreactor as shown in *Figure 6*.
3. Autoclave at 121°C, 15 p.s.i. for 30 min.

Cell and medium inoculation

Both these steps are performed aseptically in a laminar air flow cabinet.

1. Pour the inoculum of cells into the glass bead reservoir.
2. Allow cells to adhere overnight in a 37°C hot room.

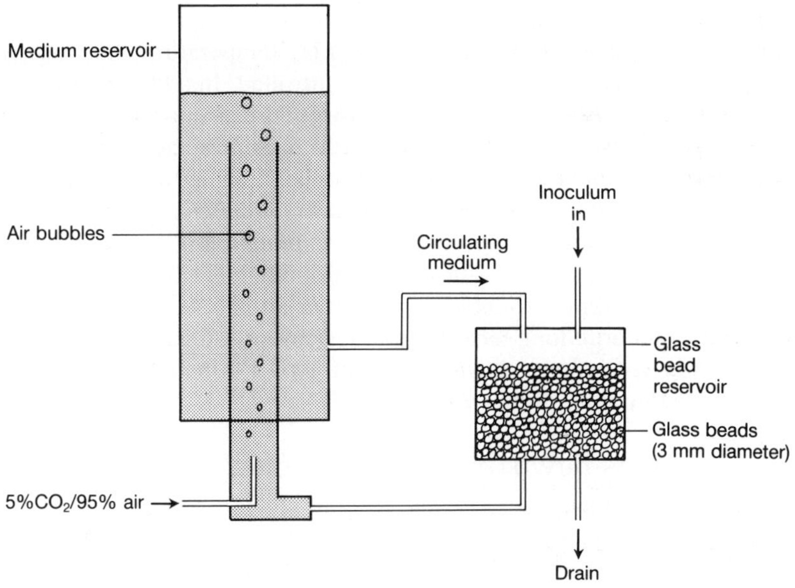

Figure 6. One litre glass bead propagator for batch cultivation of anchorage-dependent cells. (Reproduced with kind permission from R. Brydges.)

Protocol 2. *continued*

3. Pour the serum-supplemented medium into the medium reservoir.
4. Transfer to a 37°C hot room and re-circulate aerated medium into the glass sphere propagator.
5. Incubate for 3–4 days.

Recovery of cells by trypsinization

1. In the laminar air flow hood, drain the spent culture medium into a 1 litre Duran bottle.
2. Add 200 ml of EDTA/trypsin solution to the glass sphere reservoir.
3. Incubate at 37°C for 15–30 min, shaking the reservoir intermittently.
4. Pour the detached cell suspension and beads into a sterile beaker.
5. Centrifuge at 1500 r.p.m. for 10 min.
6. Re-suspend the cell pellets into fresh serum supplemented media at a final density of 5×10^4 cells/ml.

3. Long-term continuous cultivation

An alternative approach to batch cultivation is to continuously add fresh medium to the cells and to remove either medium mixed with cells, or cell-free medium from the bioreactor. A well-designed system should aim to maintain the cells in a viable state for prolonged periods, lasting up to several months. As in batch cultivation, culture pH, temperature, and dissolved oxygen tension need to be monitored and controlled. In addition, automated pumps are required to control the addition of fresh nutrients and for the removal of wastes and products. Fluid transfer lines from media reservoirs, or leading to waste and harvest vessels must be designed for quick, sterile connection. Regular sampling for cell growth, product accumulation, substrate consumption, and metabolic waste accumulation is required. Samples are usually taken off-line, although some reliable, automated, on-line analysers may become available on the market in the near future. The maintenance of sterile long-term continuous cultivation is a major challenge in mammalian cell culture technology. The typical kinetics associated with batch and continuous cultures is shown in *Figure 7*.

3.1 Continuous-flow stirred tank reactors

In the continuous-flow, stirred tank reactor (CSTR or Chemostat), fresh medium is fed into the bioreactor at a constant rate, and medium mixed with cells leaves the bioreactor at the same rate. A fixed bioreactor volume is maintained and, ideally, the effluent stream should have the same composition as the bioreactor contents. The culture is fed with fresh medium contain-

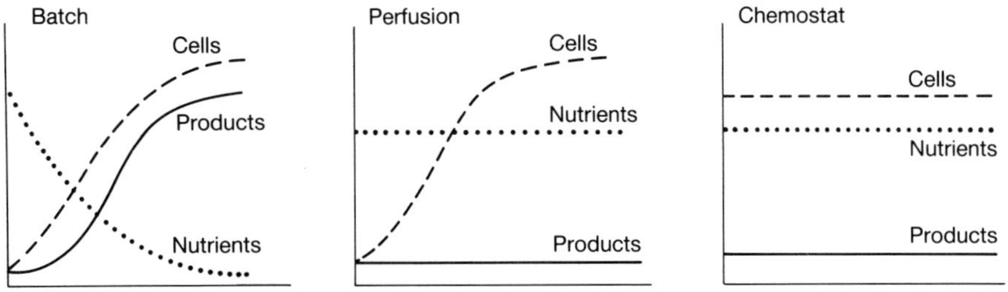

Figure 7. Kinetics of nutrient consumption and product accumulation.

ing one (and sometimes two) growth-limiting nutrient (such as glucose). The concentration of the cells in the bioreactor is controlled by the concentration of the growth-limiting nutrient. A steady-state cell concentration is reached where the cell density and the substrate concentration are constant. The cell growth rate (μ) is controlled by the dilution rate (D) of the growth-limiting nutrient.

At steady-state,

$$\mu = D \qquad (9)$$

where $D = F/V$ (F is the medium flow rate and V the culture volume). The chemostat culture offers a useful method of manipulating the culture environment for improved cell and/or product yields.

3.2 Perfusion technology

In perfusion systems, the cells are retained in the bioreactor, fresh medium is supplied continuously, and waste (always cell-free) medium is removed continuously at the same rate. Suspended mammalian cells may be entrapped within hollow fibres (10), flat membranes, or porous matrices; retained in stirred tanks by filters; or encapsulated within gels and beads (11). Anchorage-dependent cells may be cultivated on microcarriers retained in stirred tank reactors by filters; porous microspheres in fluidized bed reactors; or hollow fibre and ceramic matrices. Perfusion cultivation systems are becoming increasingly popular because they can achieve high cell densities and product concentrations, and require relatively less space and labour. A major drawback of these systems is the increased risk of contamination due to poor equipment design and lack of trained operators. However, some perfusion systems, e.g. the hollow fibre reactors for suspended hybridoma cells (12), and microcarrier cultivation for anchorage-dependent cells (see Chapter 1) are proving to be useful process-intensified production systems.

3.2.1 Hybridoma cultivation in a hollow fibre perfusion bioreactor

As an example of a perfusion system, the principle of operation of the Celltronics (New Brunswick Scientific) hollow fibre reactor will be described.

(*Protocol 3* details the experimental steps necessary to set-up this system.) The Celltronics is a laboratory-scale, semi-automated, hollow fibre bioreactor, used primarily for the cultivation of hybridoma cells secreting monoclonal antibodies. The system consists of a hollow fibre cartridge linked to a perfusion circuit which oxygenates the medium, replenishes nutrients, and removes waste and product streams. A diagrammatic representation of the perfusion circuit is shown in *Figure 8*. The hollow fibre cartridge consists of a closed shell which is packed with a bundle of porous hollow fibres. The cartridge is disposable and is supplied pre-sterilized with ethylene oxide. The pore sizes of the fibres range from 6000 to 10 000 daltons.

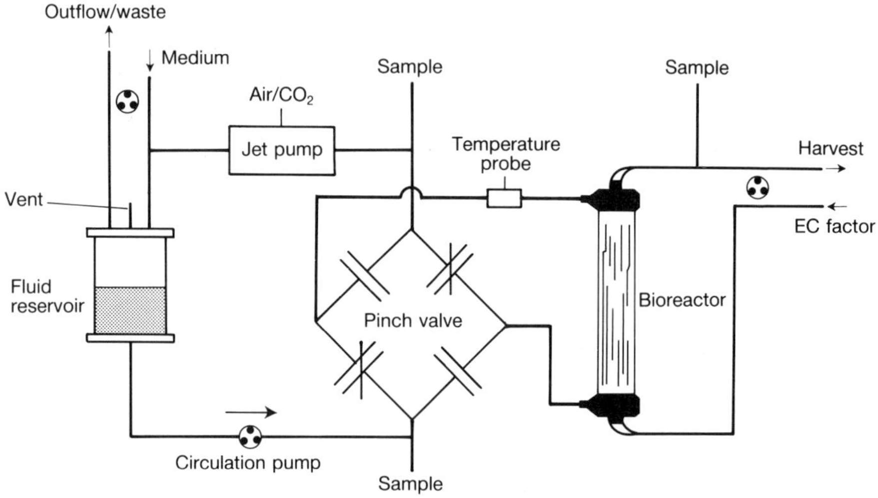

Figure 8. Perfusion circuit for the Celltronics hollow fibre bioreactor. (Reproduced with kind permission from New Brunswick Scientific.)

Cells are inoculated and cultured in the extracapillary space (EC). Addition of serum-supplemented medium or a defined serum-free medium to the EC ensures that the cells are provided with appropriate growth factors. The product (mAb) is secreted in the EC and is removed continuously via a harvest line. Oxygenated basal medium is delivered to the intracapillary space (IC) via a medium reservoir. The basal medium allows diffusion of fresh nutrients (e.g. glucose, glutamine, vitamins, etc.) to the cells, and removal of low-molecular-weight waste components (such as ammonia and lactic acid) away from the cells. A constant total volume is maintained in the system because

Basal medium addition rate to IC = waste removal rate from IC

and

Serum addition rate to EC = product removal rate from EC.

Oxygenation of the basal medium is carried out in the jet pump—a device which introduces small air bubbles into the flowing medium. Periodic reversal of the oxygenated medium to the hollow fibre cartridge is carried out by a pinch valve assembly. This medium reversal ensures that fresh nutrient supply and waste removal occur at both ends of the hollow fibre, thus reducing the build-up of concentration gradients along the length of the fibres.

Monoclonal antibody production in hollow fibre bioreactors can be closely controlled and predicted by rigorous monitoring and control of both physical and biochemical parameters. The selected optimization conditions are:

- a culture pH of 7.1 and temperature of 37°C
- a dissolved oxygen concentration >80 mm Hg
- glucose and glutamine concentrations >150 mg/dl and 2.5 mM, respectively
- lactate and ammonia concentrations <150 mg/dl and 2.5 mM, respectively
- continuous harvesting of product.

In carefully optimized systems, the mAb production rates can be seen to increase linearly with the uptake rates of glucose and glutamine (*Figure 9*).

Protocol 3. The use of the hollow fibre bioreactor

Set-up procedures

The hollow fibre cartridge is supplied pre-sterilized; however, the other components of the cultureware have to be washed, autoclaved, and aseptically connected to the cartridge. After assembly, the whole system has to be flushed with sterile water to remove traces of ethylene oxide.

1. Clean and wash all cultureware components with 7X detergent and rinse with cell-culture quality water.
2. Connect the medium reservoir, jet pump, and pinch valve assembly to the silicone tubing IC circuit.
3. Place in a sealed bag and autoclave at 121°C for 60 min.
4. Under a laminar air flow hood, aseptically connect the hollow fibre cartridge to the autoclaved cultureware.
5. Transport assembled bioreactor to the Celltronics incubator and flush 5.5 litres cell-culture water through the IC and 0.5 litres through the EC.
6. On completion of flushing, replace the water with basal medium to the IC and serum-supplemented medium to the EC.

Cytotoxicity testing

It is essential to check that all traces of ethylene oxide have been removed from the cartridge. Samples of media from the IC and EC circuits are removed and tested for their ability to sustain cell growth in small-scale, batch cultures. Trace amounts of ethylene oxide can result in slow growth or

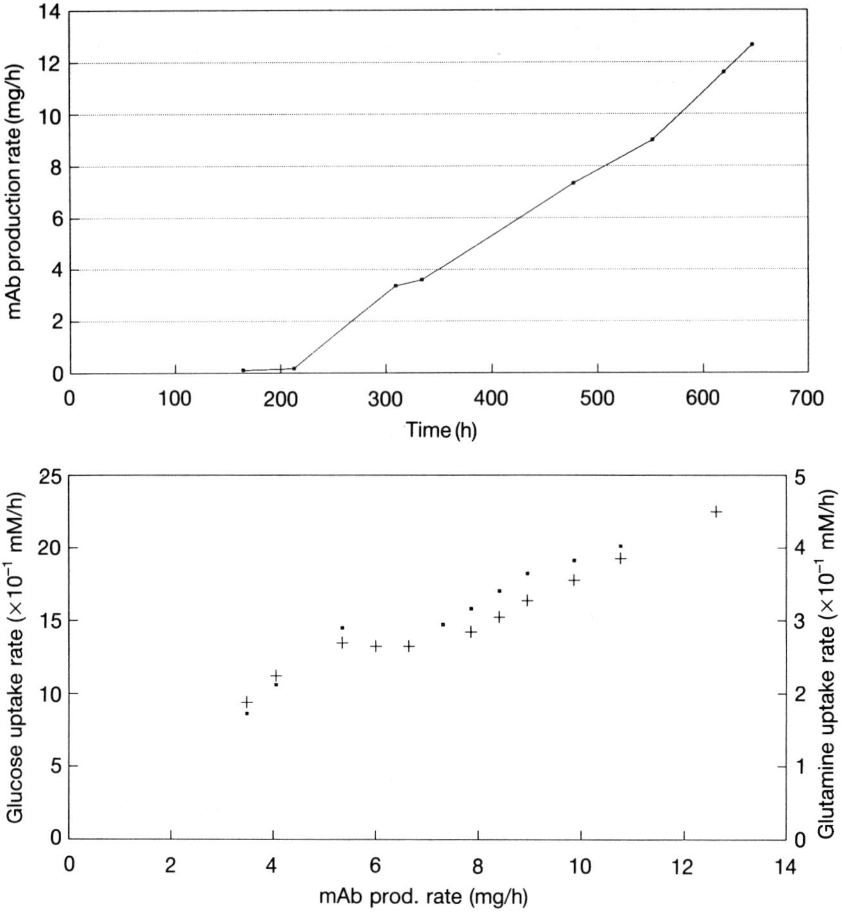

Figure 9. Monoclonal antibody production in a hollow fibre bioreactor. Key for bottom graph; •, glucose uptake rate; +, glutamine uptake rate.

Protocol 3. *continued*

instantaneous cell death. Further flushing may be required for complete removal of ethylene oxide.

1. Flush the bioreactor with cell culture media to the IC and EC for 24 h.
2. Remove 50 ml samples from the waste and harvest lines.
3. Inoculate the media samples with 1×10^5 cells/ml hybridoma cells and incubate at 37°C for 5 days.
4. Sample daily, count viable cells, and determine growth characteristics of the cells.

Inoculum preparation

A total of 2×10^8 viable cells is required for inoculation of the hollow fibre cartridge. It is important that the cells have a viability >95%.

1. Expand a hybridoma cell line in a 2 litre surface-aerated, stirred culture or in a 2 litre stirred tank bioreactor as described previously (*Protocol 1*).
2. Aliquot the cell suspension in 4×500 ml centrifuge tubes and centrifuge for 10 min at 1500 r.p.m. Decant the supernatant, and pool the cell pellets into 50 ml of fresh serum supplemented medium (e.g. RPMI supplemented with 5% fetal calf serum).
3. Pump the cells (at a rate of 200 ml/h) into the bioreactor via the harvest line.

Monitoring and controlling the cultivation

The culture is initially delivered with basal medium at 40 ml/h and serum-supplemented medium at 1 ml/h, respectively. The circulation rate of the medium through the jet-pump is initially at 100 ml/h, gradually increasing to 500 ml/h after 8–12 weeks. The medium flow is initially reversed every 30 min; this may increase as the run progresses. The rates at which media are supplied or product/wastes removed during the course of the cultivation are determined by many factors.

1. Maintain an adequate supply of oxygen to the cells by increasing the re-circulation rate and the medium flow reversal rate.
2. Maintain pH at 7.1 by adjusting the amount of CO_2 delivered to the system, or by increasing the basal medium through-put.
3. Increase the basal medium addition if glucose and glutamine concentrations fall below 150 mg/dl and 2.5 mM, respectively.
4. Increase basal medium addition, waste removal, and flow reversal when concentrations of lactate and ammonia rise above 150 mg/dl and 2.5 mM, respectively.
5. Increase the supply of serum-supplemented medium (or defined-serum free medium) only when the concentrations of product is increasing.

Daily sampling and assaying of nutrient and waste materials gives a better understanding of the cellular activity in the reactor. The relatively simple optimization strategy described above results in marked improvements in productivity and process economics.

4. Conclusion

In vitro mammalian cell culture is currently carried out in a diverse range of bioreactor designs ranging from batch air-lift and stirred tank to perfusion and continuous flow systems. At the smaller scales of operation both the conventional and the novel bioreactor designs are relatively easy to operate. At the larger scales of operation, problems of maintaining bioreactor sterility and providing an adequate oxygen supply to the cells have yet to be resolved.

Acknowledgements

I wish to thank Sue Gomm for typing the manuscript. My thanks also go to Simon Nikolay and Richard Brydges for assistance with the diagrams.

References

1. Feder, J. and Tolbert, W. R. (1983). *Sci. Amer.* **248,** 36.
2. Katinger, H. W. D. and Scheirer, W. (1982). *Acta Biotechnol.* **2,** 3.
3. Katinger, H. W. D. and Scheirer, W. (1985). In *Animal Cell Biotechnology*, Vol. 1 (ed. R. E. Spier and J. B. Griffiths), p. 167. Academic Press, London.
4. Atkinson, B. and Mavituna, F. (1983). In *Biochemical Engineering and Biotechnology Handbook*, p. 579. Macmillan, London.
5. Bailey, J. E. and Ollis, D. F. (1986). In *Biochemical Engineering Fundamentals,* p. 457. McGraw-Hill, London.
6. Moo-Young, M. and Blanch, H. W. (1987). In *Basic Biotechnology* (ed. J. Bu'lock and B. Kristiansen), p. 133. Academic Press, London.
7. Cherry, R. S. and Papoutsakis, E. T. (1990). In *Animal Cell Biotechnology*, Vol. 4 (ed. R. E. Spier and J. B. Griffiths), p. 72. Academic Press, London.
8. Handa-Corrigan, A., Emery, A. N., and Spier, R. E. (1989). *Enzyme Microb. Technol.* **11,** 230.
9. Handa-Corrigan, A. (1990). In *Animal Cell Biotechnology*, Vol. 4 (ed. R. E. Spier and J. B. Griffiths), p. 122. Academic Press, London.
10. Knazek, R. A., Gullino, P. M., Kohler, P. O., and Dedrick, R. L. (1972). *Science,* **178,** 65.
11. Griffiths, J. B. (1990). In *Animal Cell Biotechnology*, Vol. 4 (ed. R. E. Spier and J. B. Griffiths), p. 149. Academic Press, London.
12. Tyo, M. A., Bulbulian, B. J., Zaspel, B. and Murphy, T. J. (1988). In *Animal Cell Biotechnology*, Vol. 3 (ed. R. E. Spier and J. B. Griffiths), p. 357. Academic Press, London.

8

Monitoring and control of bioreactors

DOUGLAS G. KILBURN

1. Introduction

The increased commercial demand for proteins produced in animal cell cultures has highlighted the need for better, more economical production methods. Experience in the fermentation industry indicates that the performance of a biological process can often be vastly improved by a better understanding of the influence of environmental factors on cell productivity. A considerable expertise has been built up on methods to regulate the culture environment. These methods have been applied to mammalian culture with considerable success. New methods have been develped for high-intensity culture which achieve cell concentrations in bioreactors up to two orders of magnitude higher than the maximum seen in flask cultures. These high cell concentrations are totally dependent on the maintenance of a controlled culture environment. This chapter outlines the monitoring and control procedures which are most frequently used in mammalian cell bioreactors. The technology has been adapted from the process control systems commonly used for microbial cultures (*Figure 1*). The terminology used to describe these standard methods of process control is summarized in *Table 1*.

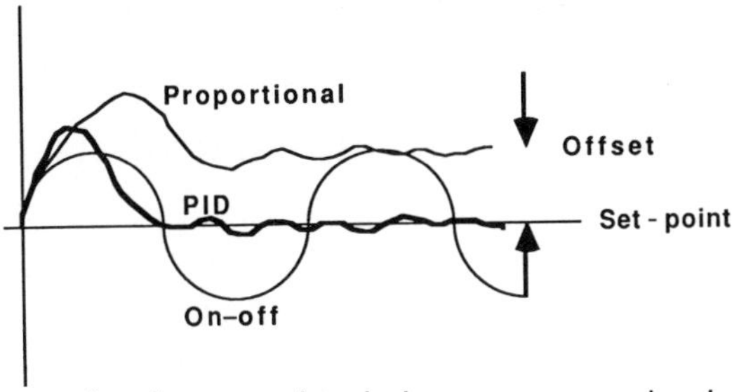

Figure 1. Process control performance. A typical response curve showing the influence of the control mode on the regulation of the controlled variable. On–off control is intrinsically cyclic. Proportional control reduces these fluctuations but the control point is always offset from the set-point. Proportional plus integral and derivative control (PID) eliminates offset and decreases the response time for control action.

Table 1. Process control terminology

- *On-off control.* In this mode the control action is either fully on or fully off. The response of the measured variable oscillates around the set point. It is best suited to steady-state conditions where large sudden deviations from the set-point do not occur.
- *Proportional control.* In this control mode the output of the controller is proportional to the error signal, such that

  $$\Theta = \Theta_o + kE$$

  where Θ is the output signal, Θ_o the output signal when there is zero error, k is the gain, and E is the error.
- The controller gain k is often specified in terms of the *proportional band.* The proportional band is the error (as a percentage of the control range) required for 100% control action. Under proportional control the controlled variable will normally be offset from the set-point (see *Figure 1*). With a very narrow proportional band the control is basically on–off.
- *Integral control.* In the integral control mode the output of the controller is a function of the integral of the error with time t. Control action increases with time as long as an error is registered.

  $$\Theta = \Theta_o + \frac{1}{t_i} \int E dt$$

  where t_i = integral time (constant). Integral control is seldom used alone but usually in conjunction with proportional control to eliminate the offset inherent in the latter control mode.
- *Derivative control.* In the derivative control mode the output of the controller is proportional to the rate of change (derivative) of the error,

  $$\Theta = \Theta_o + t_d \frac{dE}{dt}$$

  where t_d is derivative time (constant). The derivative mode is never used alone. Its function is to accelerate proportional control action when the error is changing rapidly.
- *Three-term control (PID).* This is the combination of proportional plus integral plus derivative control. When suitably tuned by adjusting the proportional band, integral time, and derivative time, PID control provides the optimum degree of process control.

2. On-line measurement and control systems

2.1 Temperature

Control of temperature in bioreactors is a well established technology and precision levels of $\pm 0.5\,°C$ or better are routinely achieved. Several different types of sensor may be used to monitor temperature.

(a) *Mercury-in-glass thermometers* are frequently used as a check on the temperature in small bench-scale reactors. The thermometer is normally inserted into a stainless steel well fixed in the head plate or side of the vessel and filled with a non-volatile fluid.

(b) *Mercury contact thermometers* are often used to control the temperature of a circulating water bath used with small bench-scale reactors. When the contact thermometer is located in the water bath, the actual temperature in the bioreactor should be checked and the set-point of the contact thermometer adjusted to give the required culture temperature in the bioreactor.

(c) *Electric-resistance thermometers* are the most frequently used sensors for temperature control in bioreactors. The sensing element is a fine nickel or platinum wire whose resistance, nominally of about 100 ohms, varies with temperature resistance. These thermometers have the advantages of stability, sensitivity, accuracy (±0.25%), and linearity of response. They respond rapidly to temperature changes and can be located remotely from the controller. The thermometer consists of a sensing element in a small-bore stainless steel tube inserted directly into the vessel or used in a thermometer well. The latter arrangement is more convenient for small reactors which are sterilized by autoclaving.

(d) *Thermistors* are semiconductors which exhibit a large change in resistance with a small temperature change. They are relatively cheap and give stable reproducible results. The major disadvantage of thermistors is lack of linearity. Their use is usually restricted to small bench-scale vessels.

The control of temperature in animal cell bioreactors is easier than in microbial fermenters because of the lower level of metabolic activity. Several control techniques are used.

(a) Temperature-controlled water can be circulated through the jacket of the vessel to control the culture temperature. Many bench-scale and most pilot-scale or larger vessels use this approach in conjunction with a control system which regulates the temperature of the water entering the jacket.

(b) In small bioreactor systems the temperature can be maintained by placing the reactor in an incubator or by controlling the temperature of air blown across the reactor. This system has the advantage that the water connections do not have to be broken to move the vessel, e.g. into a hood for inoculation.

(c) In many small bioreactor systems electrical heating of a pad in contact with the vessel is used to maintain the culture temperature. The temperature control is normally proportional or three-term using a sensor immersed in the medium. On–off control is possible if precise control of temperature is not essential. However, microprocessor controlled-temperature regulators featuring full control capability are relatively inexpensive. For small vessels this type of temperature control is simple, convenient, and precise (better than ±0.5°C). It is our method of choice for bioreactors that are sterilized by autoclaving. The reactors themselves are cheaper and less complex than jacket vessels and less likely to suffer breakage during handling.

2.2 Agitation

The rate of rotation of the agitator in a stirred-tank bioreactor is an important operating parameter, but in animal cell culture it is seldom modulated for control purposes. Animal cells are sensitive to liquid shear forces. This limits the maximum impellor tip speeds that can be applied. In normal practice the best stirrer speed for a given application is established empirically and held constant throughout the culture. Values ranging from 50 r.p.m. to a maximum 200 r.p.m. are common. The tachometers used to monitor r.p.m. generally utilize standard electromagnetic induction techniques and require little attention.

2.3 Flow rate

2.3.1 Gas flow measurement

The most common method for measuring gas flows into or out of a bioreactor is by means of a variable area flow meter, or rotameter. The rotameter consists of a vertically mounted glass tube with a tapered bore containing a freely moving ball or float. The float rises to the point where the buoyant drag forces are balanced by the force of gravity. The position of the float in the tube determines the flow area and indicates the flow. In these simple glass rotameters there is no electrical read-out so the flow measurement cannot be coupled to a control system. The accuracy is usually about ±2% of full scale over a 10:1 operating range. The float position in the rotameter is a function of the gas density. The indicated flow is thus dependent on the gas composition, pressure, and temperature. Rotameters provide a simple, inexpensive means to monitor gas flows on a bioreactor. They are normally installed upstream from sterile filters and hence are not sterilized.

Electronic mass flow meters are now becoming common for measuring and controlling gas flows on bioreactors. One type of instrument utilizes the thermal conductivity of the gas. The great advantage of mass flow measurement is its accuracy and its independence of pressure or temperature. In contrast, if a rotameter is used, provision must be made to account for the local temperature and pressure. Combined mass flow meter/controllers provide a means of measuring and automatically controlling the weight flow rate of gas.

The high accuracy and stability of mass flow meters (±1% of full scale) is particularly important for the mass balance calculations used to determine oxygen uptake rates (see Section 2.4.4).

2.3.2 Liquid flow measurement

Measurements, such as medium supply rates or alkali addition rates to a bioreactor, pose a greater problem than the gas flow measurements just discussed. The flow system and the liquids themselves must be pre-sterilized and the flow metered without danger of introducing contamination. It is possible to insert sterilizable rotameters or electronic flow meters into the lines but this is seldom done with laboratory or pilot-scale equipment. A simpler and cheaper alternative is to use indirect methods.

(a) The supply vessel can be mounted on the platform of an electronic balance or load cell. These devices provide a continuous read-out of the weight of liquid remaining in the vessel. In most cases a peristaltic pump is used to deliver the liquid from the supply vessel to the bioreactor. This pump can be activated, as required, by a controller, e.g. to add alkali for pH control. Alternatively, the rate of pumping can be regulated by a computer to maintain a preset rate of supply vessel weight change, e.g. for perfusion of medium.

(b) An alternate method is to use a calibrated metering pump to transfer liquid at a preset rate. The problem here is to find an accurate, stable metering pump capable of aseptic operation. We have yet to find a pump that really satisfies these requirements. Either peristaltic or diaphragm pumps can be used for this purpose but provision should be made for recalibration (e.g. *Figure 2*). When peristaltic pumps are in continuous operation the pump tubing suffers wear. Not only does this alter the calibration of the pump, it can result in tubing failure which will terminate the culture. A sufficient length of tubing should be allowed so that

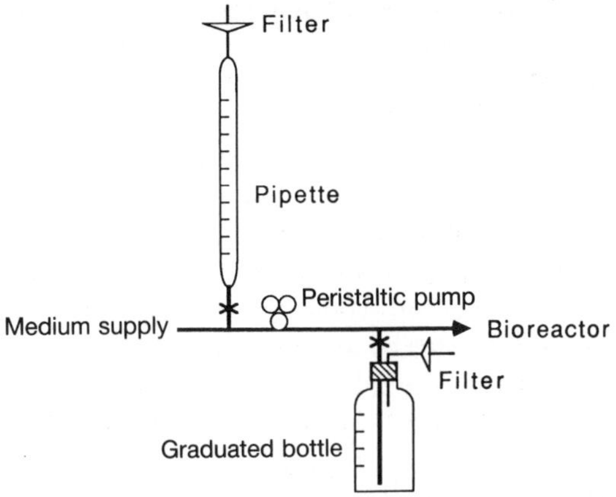

Figure 2. A system for the calibration of liquid flow under sterile conditions. The pumping rate can be measured by clamping off the main supply and timing the flow of fluid from the pipette. Alternatively, the line to the bioreactor can be clamped off downstream from the graduated bottle and the flow measured by allowing it to accumulate in the graduated bottle.

when wear becomes evident the tubing can be moved to alter the portion of the tube in contact with the pump rollers. Special tubing materials are available which are resistant to this type of wear.

2.4 Dissolved oxygen

Arterial blood has a pO_2 of about 100 mm Hg which decreases to about 40 mm Hg in the venous return. Values of pO_2 measured in tissue range from around 30 mm Hg to less than 5 mm Hg. Thus, *in vivo* most cells would never see a pO_2 greater than about 60% of air saturation (at air saturation the $pO_2 = 159$ mm Hg). One would thus expect that the optimum dissolved oxygen partial pressure for animal cell culture would be substantially less than the air saturation value. High oxygen tensions are undoubtedly toxic for animal cells, although the actual pO_2 at which this toxicity is manifest appears to vary between cell types. A number of studies using cell lines have shown a broad optimum for pO_2 ranging from almost 100% down to about 25% of air saturation (1). These values vary from cell to cell with some types being more fastidious in their oxygen requirements than others. In the course of their adaptation to growth *in vitro*, cell lines have altered markedly from their ancestors. Their adaptation may include a fundamental change in oxygen requirements or the selection of cells suited to growth under the conditions of oxygenation encountered in tissue culture. This may be particularly applicable to cell lines such as BHK and CHO which have long histories in culture. Newly selected clones or freshly isolated cell lines may have more strict O_2 needs.

Oxygen is an important nutrient for animal cells. Its availability, dissolved pO_2, is thus an important parameter in cell culture. This is quite amenable to accurate measurement and control.

2.4.1 pO_2 measurement: theory

The electrochemical basis of polarographic dissolved O_2 measurement is presented in *Figure 3*. When a suitable potential is applied between two electrodes immersed in a dilute salt solution, O_2 is reduced. The reactions for a platinum cathode and a silver anode are shown. The reduction of O_2 is dependent on the voltage driving the reaction. Above -0.5 volts, O_2 is reduced and current flows through the circuit rising to a plateau value as the voltage is increased. The plateau current is controlled by the diffusion of oxygen to the electrode and is dependent upon the bulk concentration of dissolved oxygen. Further increases in the potential between the electrodes results in other reduction reactions, e.g. at about -0.8 volts the reduction of H_2O to yield hydrogen begins.

Historically, bare metal electrodes were used to measure dissolved oxygen. However, exposure of the electrode to the sample solution leads to poisoning of the cathode surface decreasing the output. In addition, soluble components of the medium other than oxygen can be reduced. Bare electrodes are also

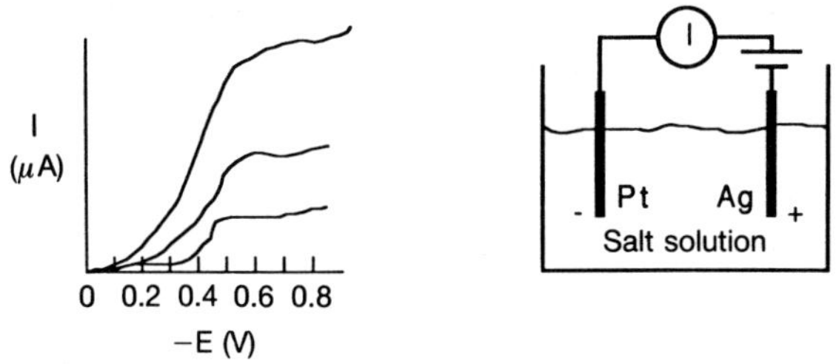

Pt cathode $O_2 + 2H_2O + 4e^- \longrightarrow 4OH^-$

Ag anode $4Ag + 4Cl^- \longrightarrow 4AgCl + 4e^-$

Figure 3. Polarographic measurement of dissolved oxygen. Current flow in the circuit on the right is limited by the rate of diffusion of oxygen to the Pt cathode. The polarograms on the left show the variation of current with voltage for three different dissolved oxygen concentrations. The current at the plateau level (i.e. at an applied potential difference of approximately −0.7 V) is proportional to the dissolved oxygen concentration.

very sensitive to fluid flow conditions near the electrode because agitation influences the boundary layer diffusion distance between the electrode and the bulk of the solution.

These problems of the bare polarographic electrode were essentially solved by the development of the membrane-covered electrode by Clark. As shown in *Figure 4*, a gas-permeable membrane is stretched across the top of the electrode. The membrane is separated from the cathode by a thin film of electrolyte. The membrane isolates the electrode from the medium and pro-

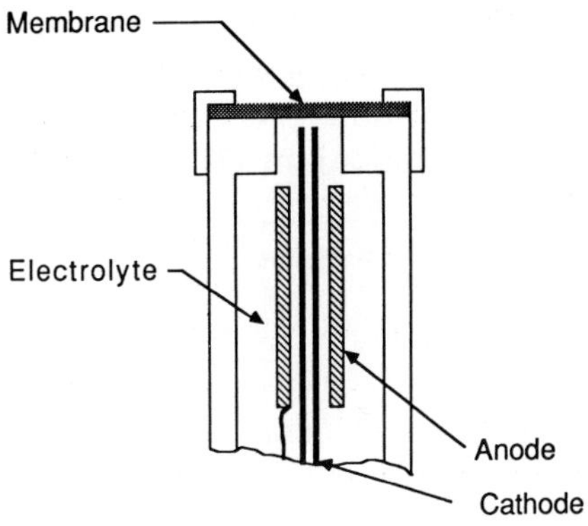

Figure 4. Membrane-covered oxygen electrode. Oxygen diffusing through the gas permeable membrane is reduced at the cathode. The current flow between the anode and cathode is proportional to the rate of oxygen diffusing through the membrane. This rate is proportional to the pO_2 in the fluid at the outside surface of the membrane.

vides a fixed diffusion barrier between the cathode and the sample. The current output of the electrode is a function of the permeability of the membrane and the partial pressure of dissolved oxygen, pO_2. The ideal electrode provides a linear output directly proportional to pO_2 according to the equation

$$i = \frac{k\,ADS}{Z}\,pO_2 \tag{1}$$

Where i is the current, k a constant, A the surface area of the cathode, D the diffusion coefficient of oxygen in the membrane, S the solubility coefficient of oxygen in the membrane, and Z the membrane thickness.

Equation 1 assumes that the entire resistance to diffusion of oxygen is in the membrane. Such an electrode would have an identical output in agitated or static medium. This complete independence of the output of the surrounding flow regime is usually not strictly true for an operational electrode. The influence of agitation rate in a stirred reactor on electrode output is dependent on the properties of the membrane. The product of the diffusion coefficient and the solubility of the gas is the permeability coefficient of a membrane (i.e. $P = DS$). *Table 2* provides data for some common membrane polymers.

The membrane-covered oxygen electrode measures oxygen partial pressure, pO_2, rather than oxygen concentration. The pO_2 is by definition the oxygen pressure that a solution exerts upon its surroundings. According to Henry's law, the solubility of oxygen in a solution is directly proportional to its pO_2

$$[O_2] = SpO_2 \tag{2}$$

where S is the solublity coefficient and $[O_2]$ is the dissolved O_2 concentration.

The solubility coefficient in turn depends upon the temperature and salt content of the solution. If one considers two solutions at the same temperature and pO_2, but with different salt concentrations, it is apparent that their oxygen concentrations will be different. For instance, at room temperature and $pO_2 = 0.21$ atm (air saturation) the O_2 concentration of water is 9.2 mg/

Table 2. Permeability of membrane materials

Membrane	Permeability[a]	
	O_2	CO_2
Silicone	804	4200
Teflon	7.6	21
Polypropylene	1.3	4.5
Mylar	0.025	0.046

[a] Permeability $\times 10^{10}$ mol/sec for 1 mm membranes, 1 cm^2 area, 20°C, 1 atm differential pressure across the membrane. Data taken from Reference 2.

litre, while that of 4 M KCl is 2 mg/litre. An oxygen electrode immersed in each solution would give the same output. The driving force that forces O_2 into the membrane of the electrode is the pO_2. The pO_2 at the cathode is virtually zero, creating a gradient for the mass transfer of O_2 from the outer surface of the membrane to the cathode. The movement of O_2 across the membrane is analogous to the transfer of O_2 into a cell. In both cases the motive force is the pO_2 (activity of O_2) in the medium rather than its concentration.

2.4.2 pO_2 electrodes

It is important to emphasize that the membrane-covered oxygen electrode is an amperometric device. In contrast, a pH electrode or redox electrode is potentiometric. The potential developed by a potentiometric electrode is an equilibrium measurement not involving the consumption of the chemical species measured. The performance of potentiometric electrodes is essentially independent of their dimensions and geometry. In the membrane-covered oxygen electrode, however, the analyte, oxygen, is consumed by the electrode. The output current is directly related to the rate of oxygen reduction. The dimensions of the electrode and properties of the membrane have a marked effect on its output current.

There are two classes of membrane-covered electrodes termed polarographic or galvanic. There is no difference in principle between the two. In the polarographic electrode, the operating potential is supplied by an external source (in the electronics of the O_2 meter) while, in the galvanic electrode, the potential is generated internally by the electrochemical couple between two dissimilar metal electrodes, e.g. a Ag cathode and a Pb cathode. Significant differences in performance can exist between these two classes of pO_2 electrode. Galvanic electrodes are usually designed to produce sufficient current output to measure directly without amplification using an ammeter. This requires high levels of oxygen consumption which often adversely affect the flow sensitivity and long-term stability of the electrode (3).

The output from an oxygen electrode exposed to a given pO_2 is increased by increasing the cathode area, increasing the membrane permeability, or decreasing the membrane thickness. High O_2 consumption rates extend the oxygen concentration gradient beyond the limits of the membrane creating a strong dependence on stirring rate.

The response time of a pO_2 electrode is primarily a function of the membrane thickness and permeability. A thin, highly permeable membrane provides the fastest response time. Aside from increasing the flow dependence, thin membranes are less robust, subject to pin-hole leaks, and are more affected by deformation during sterilization than thicker membranes. There obviously is no ideal combination of all of these factors and compromises must be made in the design of a practical electrode. Typical response times are on the order of 1 min for 90% response.

Another consideration of importance in the performance of an electrode is the current in the absence of O_2. The major source of this zero current is diffusion of O_2 from the bulk of the electrolyte to the cathode. This can be minimized by designing the electrode with a long diffusion path to the bulk of the electrolyte and a very thin film of electrolyte between the membrane and the cathode. Zero current is usually less than 1% of the air saturation current.

The O_2 permeability of polymer membranes increases exponentially with temperature. The output of a membrane-covered electrode thus has a very high temperature coefficient, usually in the range of 2–3% per degree centigrade. Since the temperature of bioreactors is normally regulated to within a fraction of a degree centigrade, temperature compensation of O_2 electrodes is not required. It is essential, of course, to calibrate the electrode at the operating temperature.

2.4.3 Calibration of pO_2 electrodes

Sterilization of a pO_2 electrode usually alters its calibration, probably as a result of stretching and thinning of the membrane. For this reason calibration is performed after sterilization prior to inoculation. The vessel should be flushed with N_2 until a stable zero current is established. The vessel is then flushed with air (often by continuous sparging) until the output stabilizes at the air saturation value. This output is nominally set as 100% (equivalent to 0.209 atm or 159 mm Hg). In short cultures of 7–10 days duration, it is usually sufficient to check the calibration at the end of the culture (water can be used in the vessel) for any drift over the course of the culture.

In long-term cultures, of months duration, a significant calibration drift can be expected. Re-calibration is therefore necessary during the course of the culture. Several approaches are possible.

(a) Calibration can be based on the measurement of the pO_2 of a sample withdrawn from a bioreactor. This method has inherent problems. Pick-up of O_2 by samples at low pO_2 can lead to serious overestimation of pO_2. This can be minimized by the use of special sampling techniques designed to minimize O_2 pick-up, e.g. the use of an injection diaphragm in direct contact with the culture medium and a hypodermic syringe. Such systems increase the hazard of contamination. A more pragmatic approach to minimize O_2 pick-up problems is to temporarily raise the pO_2 in the bioreactor to a value closer to air saturation.

O_2 uptake by cells in the sample can cause an underestimation of reactor pO_2. At moderate cell concentration, the rate of pO_2 depletion is quite slow, e.g. 10^6 cells/ml require about 1 h to utilize all of the O_2 in a sample initially at air saturation. Under these conditions it is relatively easy to extrapolate to the bioreactor pO_2 from a value of pO_2 measured with an external pO_2 electrode. A blood gas analyser or O_2 electrode respirometer can be used for this purpose. This approach becomes less reliable when the cell concentration in the sample is one or two orders of magnitude higher, as is common in high-intensity cultures. Sampling systems or recycle loops have been designed to specifically address this problem, e.g. see *Figure 5*. Such a system can serve a dual purpose, measuring the rate of oxygen metabolism (see Section 2.4.3) in addition to providing calibration data for the pO_2 electrode used for process control.

(b) If an O_2 gas analyser is available, a simple pO_2 calibration system can be installed utilizing a length of fine bore silicone rubber tubing immersed in the culture as shown in *Figure 6*. A carrier gas, e.g. O_2-free N_2, passed

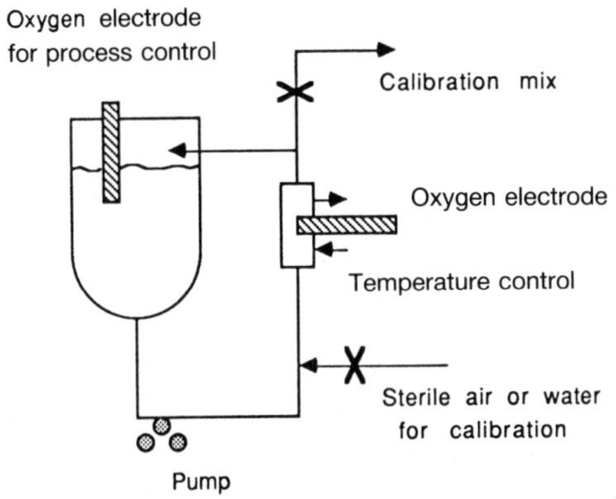

Figure 5. A recycle loop for the re-calibration of process control oxygen electrodes or the measurement of oxygen uptake rate. Culture fluid is pumped from the bioreactor through the loop at a fixed rate. The difference in pO_2 between the internal (process control) oxygen electrode and the oxygen electrode in the external loop is proportional to the pumping rate and the oxygen uptake rate of the culture. The external electrode can be calibrated by periodically exposing it to a known pO_2 gas or liquid as shown.

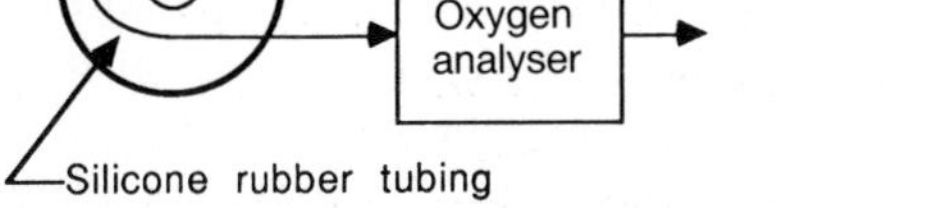

Figure 6. Dissolved oxygen measurement by the tubing method. Oxygen from the culture diffuses through the wall of the silicone rubber tube and is picked up by the carrier gas. The oxygen concentration in the carrier gas is proportional to the pO_2 difference across the wall of the tubing. A calibration curve can be constructed relating the oxygen concentration in the carrier gas, for a given flow, to the culture pO_2. Measurements of the pO_2 using the tubing probe can be used to re-calibrate the process control oxygen electrode.

through the silicone rubber tubing will pick up the O_2 which diffuses through the wall of the tubing. The O_2 content of the carrier gas is proportional to the partial pressure of O_2 external to the silicone tube. This value is also a function of the carrier gas flow rate, the length of tubing, and its dimensions. The same considerations of gas permeability discussed in reference to the membranes of pO_2 electrodes apply. This system was first proposed by Phillips and Johnson (3) to measure dissolved oxygen but is equally suited to the measurement of other gases, either dissolved or in the gaseous phase. Standard probes for use in conjunction with a mass spectrometer are available although we routinely make our own.

(c) Retractable housings designed for the replacement and re-sterilization of oxygen electrodes allow an electrode to be withdrawn at intervals and exposed to a liquid or gas phase of known pO_2 (e.g. air). Care must be taken to ensure that the temperature in the housing is identical to that of the culture.

2.4.4 Oxygen uptake rate determination

The oxygen uptake rate (OUR) of a culture is an important measure of cellular metabolism. It can be used for example, as the basis for changing medium, adding nutrient in fed-batch systems or harvesting the culture. An abrupt increase in OUR is one of the first indicators of contamination. Several methods are available for determining OUR.

(a) OUR can be determined off-line using a pO_2 electrode respirometer. In this type of respirometer, pO_2 decreases linearly with time until O_2 availability begins to limit OUR. OUR is proportional to the slope of the linear portion of the O_2 depletion curve. The linear range is cell line dependent but usually extends from about 100% to less than 20% of air saturation.

$$\text{OUR} = \frac{\mathrm{d}pO_2}{\mathrm{d}t} S \qquad (3)$$

where OUR is the oxygen uptake rate (mmol/litre/min), $\mathrm{d}pO_2/\mathrm{d}t$ is the measured slope, and S is the solubility constant. We have found that the value of S can be determined easily using a respirometer by comparing the change in pO_2 produced by injecting a small volume of sodium sulphite solution (5–10 μl) into the cuvette (2.5 ml) containing medium with the value found when the cuvette contained distilled water. In each case, 0.05 ml of saturated copper sulphate solution is added to the cuvette to catalyse the reaction of oxygen with sulphite.

(b) OUR can be estimated directly in the bioreactor by briefly interrupting aeration and following the decline of pO_2 with time using a membrane-covered probe. This approach is often used in microbial fermentations.

The actual decrease in O_2 concentration in the vessel reflects the combined effects of cellular respiration and mass transfer of O_2 to or from the culture. Continued mass transfer is provided from the head space gas and sparged gas (if any). Mass transfer is dramatically reduced by stopping the sparged gas flow and reducing agitation. If under these conditions the effect of mass transfer can be ignored, the slope of the plot of pO_2 decline versus time provides an estimate of OUR. This depends on: (i) the use of an O_2 electrode which is not significantly influenced by agitation; and (ii) the validity of the assumption that O_2 uptake by mass transfer is insignificant in comparison with respiration. This latter assumption is more likely to be true at high cell concentration, i.e. $>10^6$ cells/ml. A more stringent (and complex) application of this approach that does not ignore mass transfer has been reported (4).

(c) As pointed out in Section 2.4.3 a recycle system can be utilized to provide a continuous on-line measure of OUR (*Figure 5*). This depends on the use of two oxygen electrodes and measures the decline in pO_2 for a given time determined by the recycle flow rate and the volume of the line. It is important that air bubbles are not entrained in the flow so that O_2 supply by mass transfer does not occur. The calculation of OUR requires converting pO_2 values to O_2 concentrations such that

$$\mathrm{OUR} = \frac{\mathrm{dpO_2}}{\mathrm{d}t} S,$$

$$\frac{\mathrm{dpO_2}}{\mathrm{d}t} = (\mathrm{pO_2}(1) - \mathrm{pO_2}(2))\frac{\mathrm{F}}{\mathrm{V}} \tag{4}$$

where pO_2 (1) and pO_2 (2) refer to electrodes 1 and 2, respectively, F is the pumping rate, V the volume of tube, and S the solubility constant. The value of this type of system for continuous monitoring of OUR has been demonstrated in the Opticell bioreactor (5).

(d) Potentially, one of the most effective methods for determining OUR on-line in animal cell cultures involves performing an oxygen balance between the gas entering and that leaving the bioreactor (6). An accurate oxygen gas analyser is required in this technique. This technique is not yet widely applied for mammalian cell culture work due, at least in part, to the cost of this instrumentation. The simplest situation involves gassing the bioreactor with a constant flow of air +5% CO_2. The gas stream leaving the bioreactor is monitored for O_2 concentration such that

$$\mathrm{OUR} = \frac{F(19.8\text{-}x)C}{V} \tag{5}$$

where F is the gas flow rate (litre/min). It is assumed here that the respiratory quotient equals one and thus the flow in and out of the reactor are equal. V is the volume of culture (litres), x is the concentration of O_2

in exit gas (per cent), and *C* is a conversion factor converting O_2 used (in litres/min at the temperature and pressure of the measurement) to mmol O_2/litre/min. For 20°C and 1 atm pressure, $C = 0.416$.

Figure 7 shows a more complex system in which the gas composition and flow into the bioreactor are varied, e.g. to control pH and/or pO_2. Here the O_2 concentration and flow rate of the gas entering and leaving are measured. This data is logged by a computer (PC) and the calculation of OUR according to equation 5 is performed for each sampling time (e.g. once every minute) providing a continuous on-line output of OUR (7).

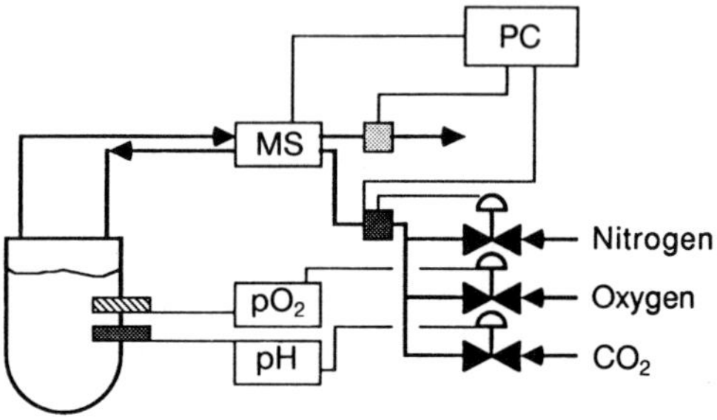

Figure 7. A system for the control of gas composition and the measurement of oxygen uptake rate. Mass flow meter/controllers are used to regulate the flow of oxygen, nitrogen, and CO_2. The oxygen and CO_2 flows are regulated automatically to control pO_2 and pH respectively. The nitrogen flow is adjusted correspondingly to maintain a constant flow rate of the mixed gases. The composition of this gas is measured before and after passing through the bioreactor by a process mass spectrometer (MS). Data on gas concentrations and flow rates are fed to a computer (PC) to calculate oxygen uptake rate, CO_2 production rate, and respiratory quotient.

2.4.5 Control strategies

A variety of techniques have been applied for the control of dissolved oxygen partial pressure in animal cell bioreactors. The complexity of these control systems vary enormously from simple on–off valves to highly automated gas-mixing systems. The current trend to significantly higher cell concentrations in cultures has increased the demands on the pO_2 control systems. The fragility of animal cells, in terms of allowable shear forces and sensitivity to bubble damage, decreases the range of control options that can be utilized. The recent development of porous microcarriers which protect cells from direct exposure to the damaging effects of bubbles or liquid shear may overcome a number of these problems.

Most hollow fibre cultures and certain proprietary systems, such as the Opticell and Verax reactors, involve the circulation of cell-free medium to a membrane aeration unit and the return of aerated medium to the bioreactor. Control of pO_2 in these reactors depends, mainly, on regulating the re-circulation rate of medium. The present discussion will not include these systems but will concentrate on the more widely used stirred-tank and air lift reactors and the ways in which pO_2 can be controlled *in situ* in these vessels.

(a) Limited control of the pO_2 of the medium can be achieved by changing the gas flow into the vessel (*Figure 8a*). This is effective in conventional air lift reactors designed to sustain maximum cell concentrations of 5×10^6 or less. At higher cell concentrations, high gas flows are required to satisfy the O_2 requirement and cell damage from bubbles becomes excessive. The inclusion of agents such as Pluronic F68 (0.01 to 0.1%) in the medium provides a degree of protection against this.

(b) Above a minimum rate, changing the gas flow into the head space of a stirred tank bioreactor has little effect on mass transfer. The application of intermittent sparging (usually of O_2) as shown in *Figure 8b* provides a simple control system for conventional systems. Cell damage from

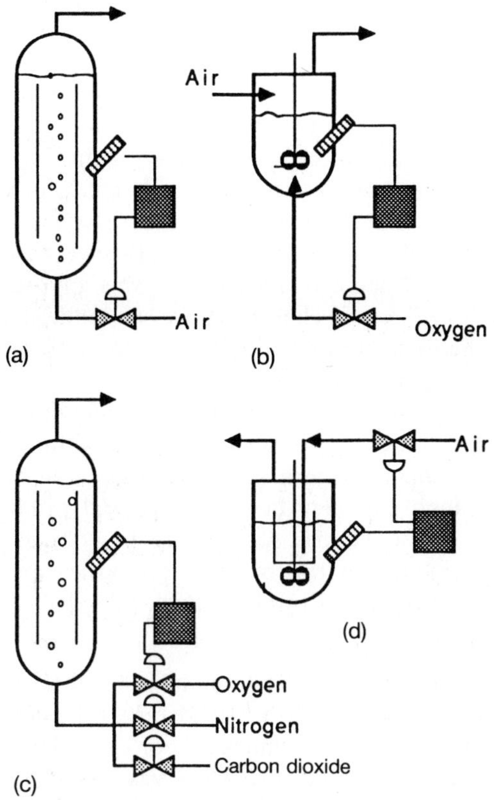

Figure 8. Strategies for the control of dissolved oxygen by: (a) alteration of air flow rate; (b) intermittent sparging of oxygen; (c) control of gas composition; (d) use of a spin filter to isolate cells from the sparge gas.

sparging limits the use of this strategy to cultures with relatively low cell concentrations. Simple on–off control is inherently cyclic. In this system the amplitude of the oscillations can be regulated by adjusting the flow of sparged O_2. Smoothing the control by reducing the sparge flow must be balanced against the limit this imposes upon the total oxygenation capacity of the system. The use of time proportional control of the sparge gas provides a better solution to this problem. Alternatively, the flow of the O_2 sparge gas can be regulated using a flow control valve. Careful balancing is still necessary here to avoid oscillation. Proportional or three-term (PID) controllers are generally used and must accommodate the relatively slow response time of the pO_2 electrode.

(c) Control of the composition of the head space gas can be used to regulate culture pO_2 in a stirred tank system. In this case a gas-mixing system such as that shown in *Figure 8c* can be used. The response times in systems such as this are very long mainly because of the lag time in the change of head gas composition and the slow mass transfer across the gas–liquid interface. Care must be taken to tune the response of the controller to avoid wide oscillations.

A similar approach of changing the gas composition can be applied to sparged aeration systems such as air lift reactors (*Figure 8c*). Sparged systems respond much more quickly than surface aeration systems to changes in gas composition and can provide much higher aeration rates. But for the problem of bubble damage, these systems would be preferred. The use of a filter screen to isolate cells from the sparge gas avoids the problem of cell damage (*Figure 8d*). This approach works well for cells on microcarriers or in clumps. However, for monodisperse cells, the mesh size of the filter must be very fine (e.g. about 10 μm) and problems with clogging are common.

An alternate means of avoiding the damaging effects of bubbles and high liquid shear rates involves growing cells within microcapsules, gel beads, or porous microcarriers. All of these immobilization systems have demonstrated high cell densities and high cell productivity over prolonged culture periods. Direct sparged aeration of these systems imposes specific requirements on the buoyant properties and strength of the particles. If these problems can be solved, the control of pO_2 should be relatively straightforward.

(d) Several bubble-free aeration systems have been tested for animal cell culture. In these systems, O_2 is provided by diffusion through a large-surface-area porous rubber membrane. For example, a length of silicone rubber tubing can be coiled inside the vessel wall (7) or even on the agitator itself. The diffusion of O_2 is proportional to the pO_2 differential or roughly to the O_2 pressure of the supply. The dissolved O_2 in the vessel can be controlled by regulating this pressure. Because the tubing area for O_2 transfer does not increase in proportion to reactor volume, these systems are limited in their range for scale-up.

2.5 pH

Control of pH is now routine in the fermentation industry. These standard monitoring and control techniques have been applied quite successfully for the cultivation of animal cells. The optimal pH range for the cultivation of these cells is within about ±0.2 pH units of neutrality. Production of lactic acid during the course of a culture depresses the pH. This can be sensed using a conventional glass combination electrode and controlled by the addition of alkali or to some extent by modifying the gas phase environment of the culture.

2.5.1 pH measurement: theory

The potential developed by a glass electrode of the type depicted in *Figure 9* can be described by the equation

$$U = E_0 - E_R - E_J + s \log a_H{}^+ \quad (6)$$

where U is the output potential which is the sum of the potentials for the circuit, E_0 is the zero potential of the glass electrode (potentials within the sealed centre glass electrode are fixed by the manufacturer and here are incorporated into E_0), E_R is the potential of the reference electrode, E_J is the junction potential between the reference electrolyte and the test solution, s is the 'slope' of the pH electrode, and $a_H{}^+$ is the hydrogen ion activity.

The pH is by definition the negative log of the hydrogen ion activity.

$$\text{Thus } U = E_0 - E_R - E_J - s\text{pH}. \quad (7)$$

Modern electrodes are designed to minimize the variation in the component potentials so that stable mV readings dependent only on the pH and the slope are obtained. The slope (mV per pH unit) is a measure of the sensitivity of the electrode and varies directly with the absolute temperature.

$$\begin{aligned} & s = 0.198\ T_K, \\ 0°\text{C} \quad & s = 54.2\ \text{mV}, \\ 25°\text{C} \quad & s = 59.2\ \text{mV}, \\ 50°\text{C} \quad & s = 64.1\ \text{mV}. \end{aligned} \quad (8)$$

The buffer in the Ag/AgCl output connection from the glass electrode is usually selected so that the potential of a new combined electrode will be 0 mV or slightly negative (−10 mV) at pH 7.0. As the electrode ages, this zero point (the pH at which $E = 0$) tends to rise (see *Table 3*). Repeated sterilization accelerates the ageing of pH electrodes. This is manifest as a decrease in slope, increase in response time, and increase in membrane resistance.

Table 3. Common causes of pH electrode malfunction and calibration drift

- pH meters are extremely high-impedance potentiometers. Poor electrical connections caused by corrosion, moisture, or insulation leakage cause erratic performance. A pH electrode stimulator can be used to check connections, cable, and electronics.
- Clogging of the reference junction can occur if medium penetrates the porous plug. Protein precipitated on contact with the electrolyte or following sterilization can block the pores causing drift and slow responses. Soaking the electrode in a protease solution may restore the response time and stability. Sulphide precipitates, e.g. Ag_2S, in the junction are evident as a blackening of the porous plug and can cause erratic performance, off-scale readings, drifting, or slow responses. Cleaning solutions are available to treat this disorder.
- Changes in the Cl^- activity of the reference electrode due to changes in the concentration of the KCl electrolyte, from evaporation or dilution by liquid entry through the reference junction, cause electrode drift. The reference cell should be flushed and filled with fresh electrolyte and the pressure balance checked to ensure a positive pressure in the reference cell at all times.
- Air in the reference cell can cause off-scale or erratic readings.
- Grounding the reference electrode to the metal body or internal components of the bioreactor results in reduction of the AgCl coating of the electrode. Under these conditions the surface of the reference wire changes from dull grey (AgCl) to bright silver (Ag) and the output of the electrode becomes unstable. To avoid this, the inputs to the pH instrumentation should always be isolated from ground.
- Repeated sterilization over a prolonged period depletes the outer gel layer of the pH-sensitive glass membrane increasing the response time. The electrode can be reactivated by soaking in a dilute HF solution.

2.5.2 pH electrodes

Virtually all pH monitoring in bioreactors now depends on the use of a combined reference and glass electrode (8). The two electrodes are arranged concentrically as shown in *Figure 9*. The reference electrode is connected to the culture by a porous ceramic junction. In conventional liquid-filled electrodes the upper portion of the reference electrode contains a reservoir for KCl electrolyte. Flow of electrolyte through the liquid junction stabilizes the junction potential and maintains the junction free of medium fouling and cellular debris. Provision must be made to prevent entry of medium through the reference junction. On pilot-scale or larger vessels, which are sterilized *in situ*, the electrode must be installed in a pressure housing so that an over-pressure can be maintained on the reference half-cell. For small vessels, which are autoclaved, the reference half-cell is not subjected to a differential pressure during sterilization. In this case it is sufficient to seal the reference during autoclaving.

During normal operation, aeration of the bioreactor creates a positive pressure. This pressure can force culture medium into the reference electrode. Medium components, especially protein or cysteine, can cause plugging of the

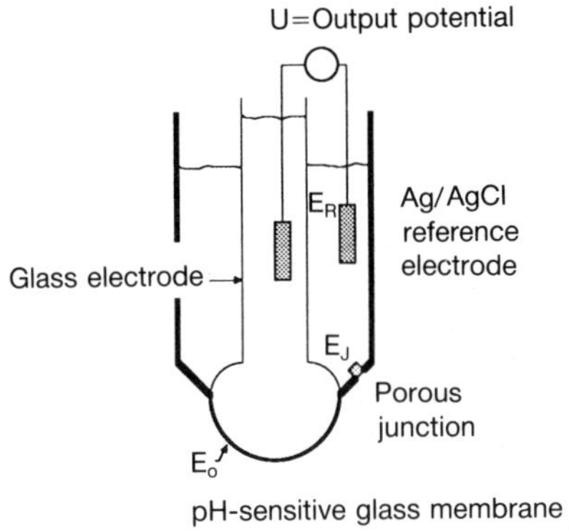

Figure 9. pH electrode. The major potentials which contribute to the output voltage (and can change as a result of fouling or ageing of the electrode) are shown: E_0, zero potential of the glass electrode under standard conditions; E_R, potential of the reference electrode; E_J, potential at the liquid junction between the reference electrolyte and the test solution, e.g. culture medium.

liquid junction to the reference electrode. This can be avoided by operating the electrode with a slight pressure (1–3 psi) above that in the vessel or by providing a sufficient height of reference electrolyte to ensure a small over-pressure of electrolyte relative to the vessel. Excessive electrolyte flow into the vessel should be avoided since this can be toxic to animal cells.

Recently, gel-filled electrodes have been introduced to obviate the need for pressurizing the reference half-cell. These electrodes are permanently sealed, in some cases under positive internal pressure. Gel-filled electrodes have a shorter lifetime, slower response time, and are less dependable than liquid-filled electrodes. However, their lower cost and convenience appear to more than compensate for these shortcomings.

Points to remember regarding the sterilization of pH electrodes are summarized in *Table 4*.

2.5.3 Control strategies

Glucose metabolism by animal cells leads to the accumulation of lactate and the depression of the pH of the culture. Even in high-intensity (high cell concentration) cultures the pH change is relatively slow and the maintenance of a constant pH is not a major process control problem. The response of pH electrodes is very rapid (typically 98% in less than 1 min) and is not limiting. Two main strategies or a combination thereof have been used to control pH in animal cell cultures.

Control of gas-phase CO_2 concentration

Most culture medium used for animal cells is buffered, at least in part, by CO_2 in equilibrium with biocarbonate

$$CO_2 + H_2O \leftrightarrow H_2CO_3 \leftrightarrow H^+ + HCO_3^-.$$

Table 4. Sterilization of pH electrodes

Sterilization by autoclaving

- Small bioreactors are autoclaved completely assembled including pH electrode, inoculation lines, sterile connectors, sample point, component filters and lines.
- The entire electrode including the electrode connection and lead must resist heat and moisture. Electrodes with an integral top connection have been designed for this application and obviate the problems caused by autoclaving leads.
- The reference half cell should be sealed to prevent loss of electrolyte. Care should be taken to unseal the reference and top up electrolyte before operation.

Sterilization *in situ*

- Pilot-scale and larger vessels are sterilized with steam *in situ*.
- Conventional, filled electrodes are pressurized to 2 atm during sterilization.
- Gel-filled electrodes need not be pressurized.
- Retractable housings are available for either conventional or filled electrodes. These housings allow a new electrode to be installed and steam-sterilized without interruption of the culture.

The gas phase in contact with the culture is usually 5% CO_2, 95% air. Decreasing the CO_2 content of the gas phase increases the pH of the culture. A simple gas-mixing system such as that shown in *Figure 10* can be used to regulate the CO_2 flow. Initially, with the CO_2 solenoid open, the air and CO_2 flows are set to give 5% CO_2 in the mixed gas. In the case of simple on–off control the CO_2 flow is maintained only as long as the pH is above the set-point. When the pH drops below the set-point, CO_2 is no longer added. As the CO_2 is diluted out of the gas in the head space of the vessel, the pH rises. The time constants in this system are long, involving a slow change in

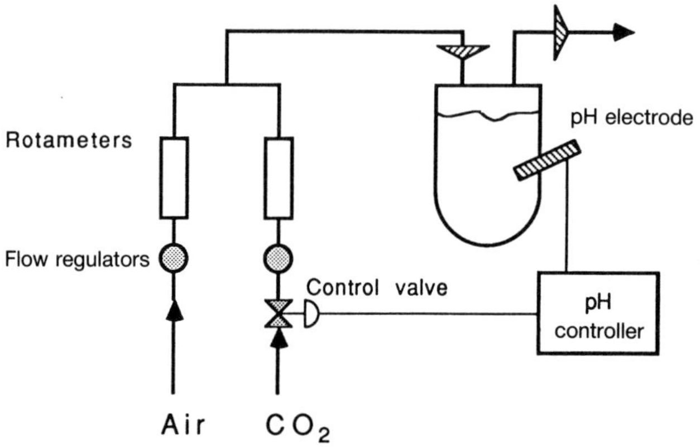

Figure 10. pH control by the regulation of the gas phase CO_2 concentration. The CO_2 flow rate is decreased if the pH of the culture drops or increased if the pH rises.

the head space CO_2 concentration and slow mass transfer of CO_2 between this gas and the medium.

Typically, in a 1 litre suspension culture of hybridoma cells with an expected final concentration of 2×10^6 cells/ml, a mixed gas flow of 100 ml/min is applied into a head space volume of 1 litre. Under these conditions it requires about 7 min to decrease the CO_2 concentration in the head space by a factor of 2, e.g. from 5 to 2.5%. The net result is a cycling pH profile. This can be minimized by using proportional control of CO_2 addition, either by time proportioning the on–off action of a solenoid, or by proportional control of a gas flow controller. The time lag associated with mass transfer of CO_2 into the liquid phase can be reduced by sparging CO_2 directly into the liquid. Aside from the potential problems of cell damage associated with sparging (see Section 2.4.4) this approach has considerable merit. The same benefit can of course be derived by sparging the entire CO_2 air mixture.

The neutralizing capacity of the 5% CO_2–HCO_3 system is in general insufficient to prevent a drop in pH late in the course of a culture. NaOH or $NaHCO_3$ must be added to the culture to bring the pH back under CO_2 control.

Control of pH by alkali addition

The alternative to control of the gas phase CO_2 content is to add alkali to the culture for pH control. We routinely use this technique for pilot-scale cultures adding either 0.65 M $NaHCO_3$ or 0.5 M NaOH. Simple on–off control is adequate here although we generally use time-proportional or three-term control of the pump on–off cycle. This technology has been well worked out and is standard practice for microbial fermentation. It has the advantage that the amount of base added can be recorded as a function of time. This is derived from the known pumping rate and a record of the pump on time. Although this technology is widely used, it has more potential for disaster than the simple CO_2 pH control system. Control failure can fill the reactor with caustic, overpressure in the vessel can prevent the addition of base, and the need to make a sterile connection to the base addition line introduces a risk of contamination. However, with careful attention to the equipment set-up and operation, this type of pH control works exceedingly well.

Late in the culture, lactate may be utilized, resulting in a rise in pH. This can be compensated for most easily by increasing the CO_2 content of the gas phase, although control by the addition of 1 M HCl is a feasible alternative.

2.6 Optional extras

Having decided on the systems to be used to control the basic environmental parameters of a culture, there is a temptation to add on a few of the extras touted by the bioreactor suppliers. It is worth discussing the pros and cons of some of these optional extras.

2.6.1 Weight/level

The value of electronic balances for monitoring or controlling fluid flow has already been alluded to (Section 2.3.2). Culture vessels may also be designed for continuous weight determination. This is of value to maintain a constant volume (weight) in the continuous perfusion systems used for high-intensity cultures. The weight controller activates a pump to add or remove medium from the reactor.

Electrical capacitance or resistance methods are also used to sense liquid level. However, the operation of these probes is confounded by foam in the vessel. Level sensors based on differential pressure or ultrasonics which do not suffer this shortcoming have recently been introduced. A simple and inexpensive alternative method which maintains a constant level in the vessel uses a pump to withdraw medium from the vessel at a rate which is always greater than the inflow. This can easily be done using a two-channel peristaltic pump with a larger tubing size on the exhaust flow.

2.6.2 Redox potential

Redox potential is a measure of the oxidation–reduction environment within the culture. The potential is usually measured using a gold or platinum redox probe with a silver/silver chloride reference electrode and employs a high impedance potentiometer identical to that used for pH measurement. It is really not clear how the redox potential measured in the culture relates to the redox state of the cell. The redox potential does vary markedly with pO_2; most of the variation occurring in the micro-aerophilic range close to zero dissolved pO_2 (9). This is below the pO_2 range of usual interest in cell culture and thus redox potential is not used as a parameter for control of the growth environment.

Redox potential can be used as an indicator of contamination in much the same way as pO_2. A sharp drop in redox potential occurs before contamination can be seen by microscope. Redox electrodes are simpler and more stable than pO_2 electrodes. They are thus very well suited to this monitoring function.

2.6.3 CO_2

Very little work has been done on the influence of pCO_2 in animal cell cultures. Most culture medium is buffered with bicarbonate in equilibrium with 5% CO_2 in the gas phase. Reducing the CO_2 concentration of the gas, for example to control the culture pH, has no obvious effect on cell growth. It appears that pCO_2 is not critical over the range of variation normally encountered in these cultures. It might become limiting if other buffer systems were used.

The partial pressure of CO_2 can be monitored using a Severinghouse electrode which measures the pH of a bicarbonate solution in equilibrium with the culture but separated from it by a membrane permeable to CO_2 but impermeable to ionic species. The measured pH is a linear function of the log pCO_2. A commercial electrode which is sterilizable has been developed by Ingold. A complex system is required to permit the filling and calibration of this electrode after sterilization. pCO_2 can also be measured using the tubing technique (Section 2.4.3) if a CO_2 gas analyser is available.

2.6.4 NADH/NADPH

The reduced forms of the nicotinamide adenine dinucleotides NADH and NADPH serve an important role as electron carriers in living cells, supplying reducing power in a number of intracellular redox reactions. The concentration of NAD(P)H in a cell culture is a function of the cell number and the level of metabolic activity. The reduced electron carriers fluoresce at 460 nm when irradiated with light at 340 nm. The oxidized forms of these molecules do not fluoresce. Sterilizable probes have recently been developed for the determination of intracellular NAD(P)H in culture systems by fluorescence. In microbial cultures the NAD(P)H concentration correlates well with cell mass during the lag phase and exponential phase of growth. This value rises during stationary phase indicating changes in the physiology of the micro-organisms. Similar results have been reported for animal cells. On-line monitoring of NAD(P)H is still at the experimental stage for animal cell cultures (10, 11). The sensitivity of the technique decreases in high-density cultures due to increased scatter and refraction of incident light (12). Nevertheless, it does provide a continuous measure of cell mass. This value is essential for the calculation of specific biological activity, for example, in conjunction with oxygen uptake rate or glucose utilization rate. Perhaps of even greater value is the potential of this method to provide information on the physiological state of the culture. Under nutrient limiting conditions, the NAD(P)H level responds immediately to changes in nutrient concentration. This response could be used to define limiting nutrients or as the basis for a control system regulating nutrient addition.

2.6.5 Gas analysis

Monitoring the composition of the gas leaving a bioreactor is a routine operation in many fermentation processes. The gas analysis is used to calculate culture parameters such as the oxygen uptake rate, the CO_2 production rate, and the respiratory quotient. The first two of these can, in turn, be related to the cell mass and provide a measure of specific biological activity. Control strategies can then be initiated based on these derived values. Although this type of process monitoring and control is still virtually unheard of for animal cell cultures, its ultimate application seems inevitable because of the potential to provide control action based on a measure of cell physiology.

In fermentation systems, CO_2 is often monitored by infra-red absorption spectroscopy. In animal cell cultures, the use of the CO_2–bicarbonate buffer system places a major constraint on CO_2 monitoring since metabolically produced CO_2 must be measured against a high background level of 5% CO_2. Unless another buffer system is adopted, the monitoring of CO_2 production will be too inaccurate to be of much value.

The use of oxygen uptake measurement has already been discussed (Section 2.4.4). The oxygen content of a gas can be determined using a paramagnetic gas analyser. Oxygen is strongly attracted to a magnetic field, a propoerty not shared by other gases likely to be found in bioreactor exhaust gas. Detectors are either mechanical or thermal with an accuracy of ±1–2 per cent and a response time of about 1 min.

A number of animal cell culture laboratories are now investigating the use of mass spectrometers for process monitoring and control. These can be either magnetic sector or quadrupole instruments. The latter are cheaper but less sensitive, less stable and less accurate than the former. The absolute accuracy of the quadrupole for O_2 in the 10–30% range is typically ±0.2% vol. with a 1% stability over 24 h. Comparable values for the magnetic sector instrument are 0.05% vol., ±0.1% relative stability. The response time of these instruments is less than 1 sec for a 90% response. Consequently, one instrument can be used to monitor a number of bioreactors (6). The input of the mass spectrometer is fitted with a sample manifold system and an automatic selector valve (typically accommodating 16 or more streams).

3. Computers

Many bioreactor systems now use microcomputers (PCs) for data logging and analysis. Their control loops are most often run locally by microprocessor controllers, each loop with a dedicated controller operating independently. The next stage of development of these computer-aided control systems involves linking the microprocessor controllers to a supervisory microcomputer which provides data logging and analysis but can also adjust the set-points on the microprocessor controllers according to some preset program or operator commands. This computer can also process on-line analytical data, e.g. oxygen uptake rates, and order control action through the microprocessor units. One microcomputer can supervise a number of bioreactor systems depending on the complexity of the systems and the amount and type of intervention required of it. Such systems are hybrids with traditional local control tied to, but operating independently of, a central computer. If the computer fails the local systems continue to run the bioreactor and the culture is not lost.

The alternative to supervisory control is direct digital control in which the computer takes over the functions of both the individual local controllers and the supervisory computer. This centralization of control requires less hardware and hence should be cheaper. However, a computer malfunction could shut down the entire system. Although direct digital control is not yet often used in conventional bench- or pilot-scale bioreactors, it has been used very successfully in specialized systems such as the Opticell bioreactor. At the moment it is difficult to make a strong case for or against direct digital control.

The role of even the most sophisticated computer controller in animal cell cultures (and fermentations in general) is relatively passive, relegated to housekeeping functions such as environment control, data logging, and the manipulation of data from on-line and off-line inputs. What is lacking here is not computing power but on-line sensor data. In most ways, the culture is still a black box. There is a fundamental need for more and better techniques for continuously sensing and analysing the physiology and the metabolic activity of the culture.

4. Off-line measurements

The focus of this chapter is on-line monitoring and control of bioreactor systems. It should be pointed out, however, that many critical variables must still be measured off-line.

- *Cell number*—determined by visual or electronic counting. Cell number is not equivalent to cell mass because the cell size varies during the course of the culture. Electronic counters (e.g. Coulter counter) determine cell volume in addition to cell number. They do not differentiate between viable and non-viable cells.
- *Glucose*—rapidly determined using an enzyme electrode based analyser. The specific rate of glucose utilization (mmol glucose per 10^6 cells/per h) is an indicator of metabolic rate. The depletion of glucose in culture medium, e.g. to 500 mg/litre or less, can be used routinely as the basis for replenishing medium or increasing its perfusion rate.
- *Lactate*—determined using an enzyme electrode-based analyser. This analysis provides information on the balance between glycolysis and oxidation of glucose. Glucose conversion to lactate is in some cases almost stoichiometric. High lactate concentration (>2.5 g/litre) has been reported to be inhibitory (12).
- *Product*—for proteins this is normally determined by an immunoassay, e.g. ELISA. Single determinations are inconvenient so samples are usually stored and assayed together providing historic data. Assays require several hours.
- *Lactate dehydrogenase (LDH)*—released from dead cells. Its activity can be measured spectrophotometrically based on the rate of NAD reduction. LDH concentration is a measure of cell death rate. This is of particular value for long-term perfusion cultures.

5. Future developments

The growth media and environmental conditions that have been adopted for large-scale animal cell culture were originally modelled on the conditions observed *in vivo*. It has been implicitly assumed that these conditions would be optimal for cells *in vitro*. Control systems have been designed to maintain homeostatic conditions of temperature, pH, and pO_2 in culture. Recent evidence suggests that this philosophy may be flawed. What is true for the whole animal may not necessarily be true for a separate part. For instance, in the case of monoclonal antibody production by hybridoma cells the optimal levels of both pH and pO_2 were lower for antibody production than for cell growth. Cell growth was optimal in the '*in vivo*' range. On the basis of this sort of evidence, it has been speculated that specific protein production may be enhanced by a certain level of environmental (or nutritional?) stress. Whether this is true or not, it indicates a change in outlook. Future developments in cell monitoring and control will depend more on an analysis of the critical factors involved in maximizing the production of specific proteins. Central amongst these will be biological factors such as nutrient flux rates, mRNA levels and stability, and the internal concentration of specific protein and its export rate.

Better probes and instrumentation will be necessary to monitor these biologically relevant parameters. For some years, it has been thought that biosensors could be designed to measure such parameters and perhaps provide a continuous read-out—a dynamic fingerprint of cellular metabolism. In the short term, it seems unlikely that biosensors, at least in the form of bioreactor probes, will provide the answer. There are too many technical problems. Sterilization of antibody- or enzyme-based probes is difficult without inactivating the sensor. In addition, their long-term stability is usually so poor that very frequent re-calibration (measured in hours or less) is essential. This is obviously impractical inside a bioreactor. These problems can be overcome by withdrawing samples from the vessels using a continuous sampling system. The sample can then be processed, e.g. using flow injection techniques, for discrete analyses or continuous measurements using biosensors. Re-calibration can be done at frequent intervals by injecting standards. Such systems are not without problems; however, in principle they do have the capacity to provide continuous data on culture physiology. Developments in this area should have a marked impact on cell culture.

References

1. Harbour, C., Barford, J.-P., and Low, K.-S. (1988). In *Advance in Biochemical/ Biotechnology* (ed. A. Fiechter). Springer-Verlag, Berlin.
2. Krebs, W. M. and Haddad, I. A. (1972). *Devel. Indust. Microbiol.* **13,** 113.

3. Phillips, D. H. and Johnson (1961). *J. Biochem. Microbiol. Technol. Engng* **3,** 261.
4. Fleischaker, R. J. and Sinskey, A. J. (1981). *Eur. J. appl. Microbiol. Biotechnol.* **12,** 193.
5. Pugh, G. G. (1988). *Bio/Technology* **6,** 524.
6. Buckland, B. (1985). *Bio/Technology* **3,** 982.
7. Kilburn, D. G. and Morley, M. (1972). *Biotechnol. Bioengng* **14,** 499.
8. Gary, K. (1989). *Amer. Biotech. Lab.* **7**(2), 26.
9. Wimpenny, J. W. T. (1969). *Process Biochem.* **4,** 19.
10. Leist, C., Meyer, H.-P., and Fiechter, A. (1986). *J. Biotech.* **4,** 235.
11. MacMichael, G., Armiger, W. B., Lee, J. F., and Mutharasan, R. (1987). *Biotechnol. Tech.* **1,** 213.
12. Reuveny, S., Velez, D., Macmillan, J. D., and Miller, L. (1986). *J. immunol. Methods* **86,** 53.

9

Downstream processing: protein recovery

ANDREW LYDDIATT

1. Introduction

Biochemical recovery (or downstream processing to adopt the popular sobriquet) encompasses those operations necessary to isolate and purify target product molecules from source feedstocks to specifications defined by applications. In respect of studies of cell physiology and metabolism, mammalian cell culture can be an end in itself, but the majority of technologies introduced in this volume have been developed with molecular productivity in mind. Current and near-future capabilities for native and recombinant protein production in animal cell systems have been reviewed elsewhere (1, 2). Such products will ultimately be representative of the broadest classification of protein molecules including monomeric and multimeric associations, hydrophilic and hydrophobic characters, and varying degrees of conjugation with carbohydrate, nucleic acid, lipid, and metal groups. Current and future progress in protein recovery from mammalian cell culture will thus largely depend upon universal advance in the design and implementation of biochemical downstream processing. There are few special circumstances surrounding mammalian cell culture which need elevate the molecular recovery of its products to a state of special technical status.

The only exceptions are associated with the likely scales of operation and the professional status of biochemical recovery itself. The table of contents of the present volume unintentionally reflects a view, still prevalent in modern biotechnology, that biochemical recovery should be considered (as implemented) at the end of the conventional sequence of manufacturing operations. However, many commercial practitioners would argue that solutions to problems of efficient biorecovery should dictate upstream choices of reactor, biological system, and process conditions in product manufacture. In contrast, the research practitioner typically has a more jaundiced view. Biorecovery is the tiresome interlude between exciting technical developments in cell culture and stimulating insights gained from the product application as novel molecular probes, standards, or effectors. Therein lies the problem for product manufacture from animal cells. The common scale of operations lies closer to the 1–50 litre batches achievable in ambitious research laboratories rather than the larger scales adopted by a small number of commercial manufacturers. It is not surprising that the majority attempt to adapt those laboratory procedures for biorecovery on larger scales. The poor status and understanding accorded purification in the laboratory, coupled with 'instant' and 'generic' solutions peddled by commercial equipment and consumable suppliers, will frequently hinder the objective exploitation of productive systems.

Successful biochemical recovery requires a thorough understanding of molecular and physical characteristics of products and impurities which can be manipulated to predispose selective partition of one from the other in unit operations of purification. Practical methods and protocols are too detailed and system specific to offer more than illustration of general principles in this short review. The reader is referred to the excellent volumes in the Practical Approach Series dedicated to protein purification methods and applications (3, 4). The present chapter discusses general principles of philosophy and practicality to better underpin the choice and implementation of efficient isolation strategies appropriate to the molecular characters and configurations of manufacturing operations.

2. Influential factors in process selection

Despite impressive technological advances detailed elsewhere in this volume, it is uncommon for any configuration of mammalian cell culture reactor to deliver the product in a form suited to direct end-use. Various processes of concentration and fractionation must be undertaken to convert the manufactured form into a working product. These processes are conventionally arranged in a sequence of operations of increasing resolution whose length and complexity directly reflect the purity of the end-product. Sequential processes of solid–liquid separation, cell breakage, feedstock concentration and fractionation, and product polishing are important, but practical advantage can be gained from the strategic integration of successive unit operations to the benefit of process efficiency and yield. Such integration is ultimately made possible by the system configuration of the upstream cell culture and an intimate understanding of the molecular characteristics of products and impurities.

2.1 System characteristics

In addition to advantages of process intensification, reduced media use, and improved productivity, individual reactor configurations will offer particular benefits in terms of the nature of feedstock delivery. Continuous processing may become a serious option, although quality control and the maintenance of material records in support of dedicated manufacturing operations will become more difficult. The use of membrane-limited reactors or immobilized/entrapped cells in porous supports may eliminate the need for primary solid–liquid fractionation, but risks remain of fine debris arising from cell lysis in the latter case.

Serum-based media form excellent growth and production systems for mammalian cells, but pose complex purification problems in biochemical recovery—not least through risks of batch variability in raw materials. Defined, low-protein media have many advocates (see Chapters 2 & 3) and have found success particularly in the manufacture of monoclonal antibodies (5) where their applications governs defined and relatively simple feedstocks for downstream recovery operations. However, protein-rich media have beneficial effects upon cell integrity and product stability in stirred tank reactors (6). Excess protein, despite posing problems in purification, may act beneficially as a random scavenging substrate for the variety of proteases released at cell death/lysis. This may be an important factor in stabilizing products produced and accumulated in extended batch cultures when residence times are long.

These, and other more intensive culture systems, will maximize overall product concentrations in crude feedstocks destined for purification. This will benefit the forward kinetic rates of partition systems exploiting adsorption or liquid-phase extraction (7). Simple concentration of feedstocks by ultrafiltration prior to product fractionation will also enhance the driving forces for such partition, but co-concentration of other macromolecular impurities is a complication.

The physiological pH (6–8) and ionic strength of most spent cell culture media (equivalent to 0.1 to 0.15 M NaCl) is a further consideration in system selection. It inevitably precludes direct product recovery by ion exchange adsorption, although simple dilution is an option if the product concentration is high. Other process additives such as antifoams or cell protectors may have adverse effects through the fouling of membrane or adsorbent surfaces.

2.2 Product specifications

Product location in cultured animal cells has an important bearing upon process selection (8). True extracellular products (actively or passively secreted) pose few problems, particularly when cell death is minimized in defined protein media. Intracellular products require cell disruption for product release. This is not technically difficult with animal cells and will yield a relatively concentrated extract. However, maximization of product release and the presence of cytoplasmic impurities may compromise particular purification processes. Membrane-bound products generally require detergent addition for solubilization and extractive processing. Such additives frequently interfere with biological activity, whilst their quantitative removal from purified preparations is technically difficult.

Product end-use will determine the extent of purification undertaken. Thus, ultrafiltered concentrates of hybridoma culture supernatants may serve as standard antibodies in immunochemical tests (ELISA, RIA, etc.). Indeed, the presence of protein impurities (e.g. serum proteins) may stabilize the product and prolong shelf life. Partially purified enzyme preparations, having guaranteed absence of contrary factors (natural inhibitors, competitive activities, proteases, etc.), may find application as diagnostics. In contrast, products destined for research or therapeutic application must be characterized by proven molecular homogeneity. Polishing operations at the end of a purification sequence are often solely employed to eliminate particular types of impurity from preparations (pyrogens, nucleic acids, proteases, etc.).

The molecular characteristics of protein products will determine the extent of purification possible in a particular preparation. Parameters of molecular mass and isoelectric point will, respectively, influence the performance of ultrafiltration and ion exchange adsorption procedures. The majority of currently eligible protein products (and contaminant impurities) have masses in the range 20 to 150 kd and isoelectric points between pH 4 and 7. Deviations from these values effectively enhance the ease of product purification, whilst other molecular abnormalities, such as heat stability, protease resistance, and solvent solubilities, can be readily utilized. More elegant exploitation of specific molecular recognition of products by antibodies, lectins, activity modulators, and biomimetics (native or genetically introduced) can impose profound selectivity upon simple partition procedures (adsorption or liquid extraction).

The overall recommendation is that thorough quantitation of the molecular characters of products and impurities (and their likely variation in productive culture) will pay dividends in the controlled and predictive manipulation of biorecovery operations. Such quantitation can be achieved by gel electrophoresis in native and denaturing conditions, isoelectric focusing, and batch binding experiments with ion-exchange and hydrophobic adsorbents. Assessment of solubilities and biological activities in various pH and ionic conditions in the presence or absence of detergents, chaotrophic agents, and precipitants is also necessary.

3. Conventional purification schemes—integration and compression

A generalized sequence of operations for product purification reflects the conventional approach, i.e. successive removal of particles, excess water, and unwanted impurities to yield defined product preparations. Upstream variation of reactor types and growth conditions will alter the degrees of emphasis placed upon individual operations, and may eliminate others. However, this simple sequence offers a framework for consideration of how individual unit operations may be integrated, combined, or eliminated in the face of particular purification problems.

A frequently cited paper by Bonnerjea and colleagues (9) reinforced the truism that multi-step processes have lower overall yields than those with fewer steps. Thus a four-stage process averaging a respectable 80% product recovery at each stage will yield 41% overall of the original material. This is not commercially attractive except for the highest value products. In the laboratory, multi-stage processes (6–10 steps—some repeated to enhance purity) may have overall yields as low as 10–15%. Such values raise questions concerning the representative nature of recovered end-products relative to those native materials lost in process wastes.

One justification for extended purification sequences is the conventional late deployment in fractionation of selective methods (e.g. affinity chromatography) at stages where feedstock volumes are small and uncontaminated with potential system foulants or antagonists. The expense and sensitivities of such processes are cited as justification for upstream purification steps applied as much to protect the selective fractionation as purify the desired product. It has been noted (9) that late use of selectivity underexploits the purification power of such techniques. Compression of purification sequences, made possible by a practical understanding of system molecular characteristics, will be enhanced by earlier use of bioselectivity with concomitant increase of process resolution and overall yield.

Mention should be made of the importance of temperature and sterility controls in biochemical recovery from mammalian cell culture. Both serum-based and low protein media are rich sources for microbial contamination. Most protein products will suffer temperature-related labilities in respect of conformational modifications, chemical side-reactions, and proteolytic decay. In laboratory-scale processes (1–10 litres), whole sequences can be readily undertaken at 4°C. Such a scale of operation could also be protected against bacterial contamination by sterile filtration of all resident and input feedstocks after the solid–liquid separation steps. Larger volumes of feedstock will impose signifiant refrigeration costs, and consideration might be given to allowing feedstocks to cool from the reactor to ambient temperatures during rapid process through primary separation and concentration procedures. Refrigeration and filter sterilization could then be more easily applied to smaller feedstock volumes prior to product fractionation. However, it is ultimately the final specification and end-use of products which govern process requirements.

3.1 Primary separations

Primary separation processes for extracellular products are generally focused upon the removal of cells and cell debris. This enables the presentation of clarified and solubilized product solutions to concentration and fractionation equipment with minimal particle fouling and contamination. Immobilized cell reactors, or those in which cells are entrapped in a matrix or behind a permeable membrane, will yield a ready-clarified feedstock, although problems of fine debris and/or colloidal precipitates promoted by physical and chemical modifications may require solution. Where the product is intracellular, solid–liquid separation serves to concentrate the cells and enable minimal volume cell disruption at solids loadings of 20–60% wet w/v to yield concentrated crude extracts of product.

3.1.1 Centrifugation

Intact animal cells are relatively large and dense and, in solution, have little influence upon bulk viscosities of feedstocks. Passive sedimentation is possible, but enhanced gravitational fields are usually employed. Batch centrifugation at low speeds (equivalent to 1500 *g*) in laboratory equipment (1–10 litres) is sufficient for many processs. Larger-scale processes (>100 litres) have been reported using native or flocculant conditioned feedstocks and continuous disc-stack centrifuges. Flocculated aggregates settle faster than single particles, and such treatment may also agglomerate the fine debris which escapes low-speed centrifugation. As with other reagents, the addition of polyelectrolytes, metal ions, or other materials as flocculants may have deleterious effects upon recovery operations. The effectiveness of flocculants must be judged against their status (like all additives) as system impurities which must either be tolerated in the end-product, or removed during subsequent processing.

3.1.2 Microfiltration

The alternative to centrifugation for solid–liquid separation is microfiltration (10). Suspension feedstocks are pumped at high velocity through channels limited by membranes having defined pore diameters (generally, 0.07, 0.2, or 0.45 μm). Transmembrane pressure gradients drive a flow of permeate through the membrane to a collection channel, whilst feedstock concentrates on the retentate side. Particular membrane configurations (hollow fibre, flat plate, pleated sheet, scroll) are associated with individual manufacturers. Such variations are designed to pack maximum membrane area into the smallest working space, whilst maintaining capabilities for the hydrodynamic and pressure variation which largely govern performance. Equipment is available for laboratory use (1–10 litres), generally in recirculating mode, and at much larger scales with recirculating or single-pass systems. Peristaltic pumps are recommended for mammalian cell systems to avoid cell damage.

Process scale-up should be undertaken carefully, since the hydrodynamic features of laboratory equipment are frequently modified in pilot- and production-scale equipment (11, 12). Steady-state permeate fluxes of 5–100 litre/m^2/h can be achieved at all scales, but rates vary with recirculation velocities, solid content of feedstocks, and transmembrane pressures. The chemical nature of membrane surfaces, and physical operating conditions, also influence product recoveries (10). Difficulties of pumping slurries at concentrations greater than 40–60% (wet w/v) ensure that quantitative recovery is impossible without dilution and refiltering (batch or continuous) of retentates to displace trapped solutes. Displacement washing is better suited to the recovery of clean cell suspensions for disruptive release of intracellular products. System variation, in terms of cell physiology, cell number, product concentration, and other factors, influences membrane performance in a complex and poorly understood manner not seen in centrifugation.

3.1.3 Cell disruption

Cells recovered by either centrifugation or microfiltration can (after washing) act as a source of intracellular products. Product release can be achieved by a variety of physical and biochemical means ranging from enzyme lysis to wet-milling (13). Many techniques applicable in the laboratory (freeze–thaw, sonication, french pressing, etc.) do not scale efficiently. In general, larger-scale, high-energy processes of milling and homogenization are not essential to disrupt the relatively delicate walls of animal cells. However, the efficiency of the low-energy techniques of osmotic shock, enzyme or chemical treatment, and freeze–thawing may be strongly influenced by the physiological state of harvested cells. Methods of cell disruption will be most economic (but not necessarily efficient) if conducted at a high solids content (20–60% wet w/v). Dilution of disrupted cells with suitable extraction buffer (± detergent or other solubilizers) will aid total product release. All cell disruption processes generate a feedstock requiring clarification of remaining solids. Low-energy processes may generate cell ghosts or large debris particles whose behaviour in centrifugation or microfiltration differs little from intact cells. High-energy sonication, milling, or homogenization may maximize product release but at the expense of the generation of fine cell debris which clarifies poorly. Flocculant addition and high-speed centrifugation are recommended here in both laboratory- and pilot-scale operations.

3.2 Product concentration

The purpose of product concentration is to reduce feedstock volumes and enhance partition kinetics in fractionation processes, without significant product loss. Concentration should primarily be regarded as a de-watering process, and is best achieved by ultrafiltration or precipitation. Impurity elimination is a bonus rather that the objective in these operations.

3.2.1 Ultrafiltration

Processes similar to microfiltration, using essentially identical equipment fitted with membranes having rejection characteristics based upon scales of molecular mass, can be used for de-watering clarified feedstocks (10, 14). Solutes can be concentrated against membranes with average pore diameters equivalent to globular molecules having masses of 10, 30, 100, and 300 kd. Hence, a 10 kd molecular weight cut-off (MWCO) membrane will reject molecules above that mass, and concentrate them in the retentate whilst low-molecular-weight solutes and solvents form the membrane permeate. Solutions can be concentrated 10–100 fold without change in phase, pH, or ionic strength. Buffer conditions can be changed, or salts removed, by the process of diafiltration which is analogous to displacement washing in microfiltration.

Variations in pore-size distribution, and irregular dimensions of protein solutes, limit the degree of achievable molecular fractionation. Thus, murine hybridoma supernatants are best concentrated against a 30 kd MWCO membrane if monoclonal antibody (IgG: 150 kd) loss to permeate is to be avoided. Hydrodynamic factors of feedstock re-circulation and pressure gradients across membranes are the major parameters which can be manipulated to optimize performance. Ultimately, the chemistry of the membrane surface and its interaction with solute molecules in the feedstock underpin the performance of equipment (10). Thus the contact of feedstock with defined membrane surfaces promotes the formation of a polarization layer (comprising products and impurities). This has a unique resistance to solute and solvent permeation controlled by re-circulation rates, transmembrane pressures, and physical and chemical factors associated with feedstock composition. These parameters, which strongly influence product recovery, will change with retentate concentration (14).

3.2.2 Precipitation

Processes whereby physical conditions or chemical additives can be manipulated to cause precipitation of protein products have a long history of application in biochemical recovery. Manipulation of molecular conformation by temperature, water associations by added salts, or net electrostatic charge by pH can contribute singly or in concert to these processes. Fractionation can be limited by the common behaviour exhibited by the majority of protein molecules, whilst quantitative recovery of biological activity from a crude precipitate is not guaranteed. This is not surprising since protein precipitation (particularly that physically induced by temperature or pH) closely parallels partial protein denaturation. A further problem arises from the solid–liquid separations necessitated by precipitate generation. As with lystate clarification, centrifugation performs best at both laboratory and production scales. Residues of precipitants after product dissolution may also inhibit subsequent fractionation steps. Thus residual salt would interfere with ion exchange chromatography, but could be applied to adsorption and fractionation in hydrophobic interactive chromatography.

Agents such as lower alcohols, polyethylene glycol, and sulphate salts have been widely used as precipitating agents in protein recovery (15). The manipulation in tandem of ethanol and temperature has achieved very efficient concentration (and fractionation) of blood proteins in the Cohn procedure (see Chapter 2 and Reference 16). Similar success might be predicted in serum-based cell culture media containing target products. Methanol and isopropanol have also found applications in protein precipitation, and water-soluble alcohols are generally compatible with many subsequent operations. Flame-proofing operations and instituting solvent recoveries will become important economic factors in processes above laboratory scale. Polyethylene glycol has proved an effective (if unpredictable) protein precipitant but, like flocculants, poses problems of removal in subsequent purification steps.

However, the most widely used precipitation procedures are based upon pH manipulation around the isoelectric point, or salting out phenomena with ammonium or sodium sulphate. Product concentration is more readily achieved than fractionation in both cases. Conditions of pH, product and impurity concentrations, temperature, and salt addition all have a significant bearing on process efficiency of ammonium sulphate precipitation (17). This is particularly true on the larger scales, whilst salt removal from recovered precipitates is a difficult technical problem. Corrosion of equipment on prolonged exposure to ammonium sulphate is an additional practical problem.

3.3 Product fractionation

Product fractionation should be designed to remove sufficient contaminating materials to bring the product in line with specifications of end-use. Such fractionation could range from the simple elimination of proteolytic enzymes to high-resolution dissection of natural molecular heterogeneity (isoenzyme mixtures, post-translational variants, etc.). Products having lower specifications may benefit from methods which undertake impurity subtraction rather than positive product recovery. Thus, heat treatments of stable products will eliminate labile impurities as sediments, and many precipitations with alcohols or sulphate salts can be orchestrated to leave products in solution. In the latter case, product solutions having high concentrations of precipitant may not constitute a suitable feedstock for further processing. However, many diagnostic proteins are stabilized and sold in concentrated ammonium sulphate solutions. Direct fractionation in hydrophobic chromatography is also possible with such feedstocks.

Quantifiable characteristics of electrostatic charge, surface hydrophobicity, biological affinities, and molecular size of products and impurities will underpin the success or failure of particular fractionation techniques. The common qualities of the majority of proteins (20–150 kd mass; IpH, 4 to 7, moderately hydrophilic) have already been mentioned in the context of fractionation. Characteristics of extreme isoelectric point or hydrophobicity should be identified in the initial process screening. Unique qualities will be those of natural recognition (inhibitors, substrate analogues, etc.; Reference 18), or those introduced by genetic engineering to enhance purification (19).

Many types of fractionation could be considered here. The most widely used techniques are those associated with adsorption chromatography (20). Conventional liquid–liquid extraction plays no significant role because of the incompatibility of most proteins with organic solvents. However, aqueous two-phase fractionation is currently attracting much attention (see Section 4.3). Preparative electrophoresis can be undertaken readily in the laboratory to produce gram quantities of proteins separated on the basis of molecular size or charge, and recovered in high yield and purity. Such materials have been used as diagnostics, or as antigens in the production of specific antibodies. Continuous electrophoretic devices have been designed and applied to the fractionation of mixtures such as blood plasma (21), but limited resolution and high costs have not stimulated general use.

3.3.1 General adsorption chromatography

Adsorption chromatography has been widely used for the fractionation of protein mixtures at all operational scales (3, 4, 19, 22). There are many proprietary adsorbent matrices and associated chromatographic equipment (pumps, effluent monitors, fraction collectors) marketed separately or as composite, automated kits which have demonstrable application to the high-resolution fractionation of protein mixtures produced in mammalian cell culture. In particular, there has been a move towards high-performance matrices characterized by small-diameter silica or polymer particles (5–30 μm) operated in medium- to high-pressure systems. Such developments cannot be adequately reviewed in the space available. Instead a few pertinent points, frequently omitted from review papers and commercial literature, will be considered concerning the custom design of fractionation systems.

Classical chromatography exploits differential partition of solute molecules from liquid mobile phases to the surfaces of porous solids arranged in a fixed-bed contactor. Varied residence times reflect the affinity of solutes for the solid phase and form the basis of separation in fixed-bed effluents. More widely used in protein fractionation is adsorption chromatography wherein individual proteins (or groups of closely related molecules) are adsorbed from feedstocks on to the surface of a porous solid phase under defined conditions, and are subsequently desorbed under the influence of modified liquid mobile phases (changed pH, ionic strength, polarity, etc.). The technique exploits bulk electrostatic or hydrophobic characters of molecules, as well as biospecific interactions with immobilized ligands. Solid phases may be fabricated from agarose, cellulose, acrylamide, polystyrene, silica, or composites thereof. Recent developments have favoured membrane rather than beaded geometries for solid phases. Materials are chosen for their porosity, physical strength, and biocompatibility, and are chemically derivatized with appropriate ion-exchange, hydrophobic, or biospecific ligands. Fixed-bed contactors (chromatography columns) have been conventionally used for adsorption/desorption, but adequate feedstock interactions can be achieved in a suspension reactor or liquid fluidized bed. Selectivity can be enhanced by choice of adsorbent ligand, or by modifying conditions of adsorption. There are clear advantages to such enhancement, since process efficiency will be improved if the available adsorbent capacity is devoted to product molecules and not impurities. Fractionation of bound products and impurities can be maximized by judicious selection of desorption conditions (step or linear gradients).

3.3.2 Ion exchange adsorption

This common fractionation procedure exploits electrostatic interaction between charged products in solution and oppositely charged, chemical functions covalently immobilized on porous particles (22, 23). The latter are selected for their relevant invariance of net charge across a pH range in which the bulk charge on target protein products or impurities can be manipulated. The common isoelectric points of proteins (pH 4–7) ensures that most processes operate with positively charged anionic exchangers at neutral pH values where proteins are negatively charged. Low pH operation with cationic exchangers (pH 2–5) risks product denaturation. Ion exchange adsorption is disadvantaged by the necessity for low ionic strength of feedstocks (typically equivalent to 0.01 to 0.1 M NaCl) which requires upstream desalting techniques. Desorption and product-impurity fractionation can be induced by manipulating the ionic strength (or pH) of eluting buffers in step or linear gradients. Such manipulation will impose requirements for desalting of products downstream. Close proximity of product and impurity isoelectric points will compromise the resolution of separation. Where purification is incomplete in a single step, repetition of the identical process can be as effective as the selection of additional fractionation procedures.

3.3.3 Bioselective adsorption

Replacement of immobilized ion exchange groups with molecular groups which mimic or possess native affinities for target products permits the application of bioselective adsorption processes (24). Selectivity is enhanced by the degree of specificity of interaction. When fully optimized, the procedure is strictly an adsorption operation (rather than chromatographic) since the inherent capacity of adsorbent materials is dedicated toward product association alone. Criteria for the design of solid phases, and their contact with feedstock, are similar to those detailed above. Bioselective ligands fall into two main groups. Generic ligands are those which have specific interactions with defined groups of products. Thus immobilized co-factors, or the biomimetic triazine dyes, will associate with dehydrogenase enzymes and other co-factor requiring biocatalysts. Pepsin-like proteases will associate with immobilized pepstatin, a natural inhibitor of biological activity, and microbial proteins A and G interact with various species of IgG (24). In contrast, specific ligands associate only with defined target products or their derivatives. Thus, a monoclonal antibody will normally only specifically associate with a product carrying the antigenic determinant against which the antibody was raised (25). It should be recognized that interactions which display high specific affinities are not necessarily ideal for bioselective adsorption. Monoclonal antibodies developed for ELISA diagnostic tests exhibit virtual irreversible association with target analytes and are, therefore, inappropriate for reversible bioselective adsorption. Irreversible processes might be suitable for the scavenging of specific impurities (foreign antigens, pyrogens, etc.) in polishing operations.

An advantage of bioselective adsorption which is not generally recognized, is that product–ligand interactions are robust in the face of changed physical conditions. Thus monoclonal antibody–antigen associations will sustain in a range of pH and ionic strength (typically pH 4–7 and 0.1–1 M salt; References 25 and 26). This places less constraint upon upstream processes and offers possibilities for direct recovery of products from cell culture broths—a strategy generally impossible with ion exchange adsorption. It should be emphasized that many immobilized ligands, whether small organic molecules (e.g. enzyme inhibitors) or macromolecules (e.g. monoclonal antibodies), will act as weak ion exchangers at most physiological pH values. However, dangers of non-specific, electrostatic adsorption of impurities can be minimized by routinely processing feedstocks at moderate ionic strength. Conditions for product desorption depend much upon the strength of the original interaction. It normally can be accomplished with a step change involving addition of a competing soluble ligand, of modifying ionic strength (e.g. 3 M salt) or the introduction of a chaotrophic buffer (e.g. 3 M KSCN). The two latter agents perturb molecular conformations sufficiently to promote dissociation. Exposure of products to such chemicals should be limited; thus, immediate desalting is advised. It is possible to operate a desalting column immediately downstream of the bioselective adsorbent to achieve this goal.

There are many 'ready to use' commercial products and system components which enable the practitioner to assemble a customized bioselective adsorption process (24). For example, agarose, acrylamide, and silica adsorbents carrying immobilized proteins A or G are widely available for the purification of specific IgG types produced in mammalian cell culture. Commercial bioselective adsorbents are expensive and may display unpredictable operational longevity. It has been common practice to exploit such material late in a purification process where risks of fouling or other damage are minimized. However, late use underexploits the purification power of such materials and consideration should be given to earlier use.

The relationships between solid-phase geometries, activation chemistries, ligand location and concentration, hydrodynamic factors of feedstock contact, and overall system productivity are complex and few ground rules exist for the implementation of optimized systems (26). However, the controlled assembly of bioselective adsorbents, designed for specific purification jobs, can readily be achieved at laboratory scale. Such 'in house' development will pay dividends in the advanced understanding of a separation process—a state unattainable in a lifetime of handle turning commercial systems. Individual system designs, established and validated in the laboratory, can be readily put out to tender for contract assembly when larger-scale operation is considered necessary.

3.4 Product polishing and packing

A dominant influence upon the planning of the later stages of purification of any product is the final form in which materials will be utilized, stored, or transported. Direct use as research tools or diagnostic agents generally requires a dried reagent solubilized prior to use or a stock solution in buffer compatible both with extended shelf life and/or an appropriate range of end-uses. Materials fractionated by methods discussed in Section 3.3 will typically be concentrated in strong salt or chaotrophic agents. Although some protein products are stable and can be shipped in such solutions, desalted preparations are essential for lyophilization and particular applications. Desalted preparations can be achieved by diafiltration in ultrafiltration equipment, or by processes of dialysis which are only suited to the laboratory scale.

Gel permeation (27) has not been seriously considered in this chapter as a means of macromolecular fractionation. In the laboratory, low-volume, concentrated samples can readily be fractionated in respect of molecular size on the basis of differential residence times of solutes in contact with solid phases having a defined distribution of pore diameters. However, the consequences of the necessary scale of fixed beds (typically $>20 \times$ feedstock volume) and the degrees of dilution incurred (typically 2–5$\times$) guarantees this technique a role only in the particular circumstances of specific processes (28).

The physical nature of gel permeation, wherein feedstock equilibrates and exchanges freely with the total included volumes of low-molecular-weight solutes in the mobile phase, means that the process is ideal for the removal of high concentrations of salt incurred in the elution of protein fractions from adsorbent chromatography processes. In addition, gel permeation offers an opportunity to equilibrate product feedstocks with fresh solvent systems suited specifically to enhance product dispense and lyophilization (low salt), direct product application (assay buffer), or extended shelf life (buffer plus stabilizers). The process can be run aseptically with chemical sterilization of chromatographic media and sterile filtration of all feedstocks and operational buffers. Gel permeation may also serve as a means of excluding contaminants such as pyrogens, foreign nucleic acid, and bacterial proteases which might compromise particular product specifications. With such treatments, materials can be titrated to particular concentrations and/or titres of activity, analysed for quality control records, and packaged, stored, or lyophilized in a sterile manner.

3.5 Product quality control

Quality controls will vary greatly with individual product types and projected end-use. Tests should be run on samples taken from every stage of purification in order to document the efficiency of individual stages and the yield of the overall recovery process (22). Degrees of protein purity can be assessed by native and SDS polyacrylamide gel electrophoresis augmented by the most sensitive visualization and quantification methods. Molecular heterogeneity can, in addition, be assessed by isoelectric focusing and by kinetic studies of biological activity. Some products might require validation by amino-acid sequence determination and glycosylation profiles. Determinations of specific activity (total biological activity relative to protein mass) will also illuminate the degrees of purification achieved and contribute to a definition of end product. For specialized products, the concentrations of defined contaminants will also require estimation. Concentrations of proteases, nucleic acids, adsorbent leachates, pyrogens, and microbial contaminants are all quantities which may require estimation and recording (29).

4. Process integration and control

Previous sections have purposefully concentrated upon a general account of problems posed by the design and implementation of product recovery from mammalian cell culture. Specific examples have been limited in order to avoid specialized accounts of individual unit operations. Instead, generic routes have been highlighted by which operations can be condensed, integrated, and/or eliminated in the quest for a foreshortened efficient process yielding defined products.

The present section will introduce recent ideas and developments which can potentially revolutionize the recovery of animal cell products. Specific examples are drawn from the recent research programme of the Biochemical Recovery Group, School of Chemical Engineering, University of Birmingham. Emphasis is placed upon systems associated with the recovery of monoclonal antibodies from serum-based, suspension cultures of murine hybridomas, but general principles apply to all production systems.

4.1 Integrated production and recovery of monoclonal antibodies

A recent study (30) of the controlled assembly and operation of immunoadsorbents necessitated a regular supply of monoclonal antibodies characterized by high specifications of purity and biological activity. A continuous culture facility fed through a sterile trap to a sedimentation chamber wherein the dense animal cells settled. Manipulation of dilution rates and residence times in the settling chamber generated a clarified feedstock suitable for direct contact with fixed beds of an immunoaffinity adsorbent. The latter comprised a kieselguhr–agarose composite, activated by standard cyanogen bromide procedures (26) and derivatized with human immunoglobulin G (huIgG). This protein is the specific antigen for monoclonal antibodies produced by the TB/C3 murine hybridomas (Department of Immunology, University of Birmingham).

An alternative mAb adsorbent is the protein A–sepharose composite used in Chapter 5, *Protocol 7*. The continuous culture (0.7 litre) was diluted in steady-state conditions at about 30 ml/h yielding culture supernatants post-sedimentation chamber containing approximately 50 μg/ml anti-huIgG antibody. Contaminant protein in this feedstock (3 mg/ml) comprised mainly albumin, transferrin, insulin, and cytoplasmic proteins from lysed animal cells contained in buffer solution (pH 6.5–7.0) equivalent in ionic strength to 0.15 M NaCl. Immunoadsorbents (20 ml) could be arranged on a manifold whereby a single column would load for a fixed period before substitution with a fresh adsorbent. Charged columns were back-flushed with equilibration buffer to displace fine debris and monoclonal antibody eluted in 3 M KSCN and immediately desalted on-line by passage through a Sephadex G-15 column equilibrated in phosphate-buffered saline. Monoclonal antibody products were characterized by molecular heterogeneity on SDS polyacrylamide gel electrohoresis and by high biological activity in anti-huIgG ELISA. Systems were run for up to 60 days without diminution in quality or quantity of product output.

This example demonstrates the direct use of bioselective adsorbent with a partially clarified whole broth derived from mammalian cell culture. Characteristics of the density of animal cell, the robustness of antibody–antigen interactions in the face of elevated ionic strength, and the tolerance of fine debris by relatively large composite adsorbents (450 μm average particle diameter; Reference 26) have been exploited to establish a manufacturing operation appropriate to a small diagnostic manufacturer. The scale of operation, limited here only by the economics of laboratory research, could readily be increased at least 10-fold. Labour savings compared to batch production of similar materials were substantial.

Direct bioselective recovery could be applied to all types of mammalian cell culture involving cell suspensions, entrapped cells, or immobilized cultures. Ion exchange recovery could be implemented if appropriate dilution was established. More recent work (8) with the apparatus described has indicated that attention to the cell culture in respect of cell integrity and control of temperature and residence time in the sedimentation chamber can maximize product release by hybridoma cells and yield a clear supernatant tolerated by fixed beds of smaller particles such as Sepharose and related materials (40 to 160 μm).

4.2 Fluidized bed adsorption

Discussion in Section 3 has emphasized that primary separation of cells and cell debris is conventionally a prerequisite of operations exploiting fractionation by ion exchange or bioselective adsorbent. Sedimentation of whole animal cells is not a serious practical problem. However, in feedstocks where fermentation conditions have promoted cell lysis, or in situations where cell disruption is harnessed to release intracellular products, the separation of fine cell debris (<1 μm diameter) may cause practical problems in centrifugation or microfiltration.

Liquid-fluidized bed adsorption permits the direct recovery of target product solutes from particulate feedstocks arising from whole fermentation broths or cell lysates. An increasing upward flow of feedstock through a conventional fixed bed contactor will approach a velocity which matches the sedimentation forces on individual adsorbent particles. Further increase of velocity will cause particles to separate from one another to create first an expanded and then a fluidized bed. Given a correctly sized basal sinter in the contactor, a biological suspension can be pumped through such a fluidized bed without penalty of adverse pressure drop or adsorbent blinding. Circumstances of mixing and mass transfer will facilitate all the processes of selective adsorption sought in conventional chromatographic columns. This is a relatively new approach to protein recovery, and most adsorbents developed for refined chromatographic performance are too small or insufficiently dense to fluidize efficiently. However, kieselguhr–agarose composites and polystyrene–dextran composites are two materials which perform well (31), and report has recently been made of the fluidization of homogeneous agaroses (32) albeit at lower fluid velocities.

In the simplest case, a re-circulating fluidized bed can be used as a batch adsorbent reactor for processing whole animal cell cultures or preparations of disrupted cells. A fluidized bed loop run in conjunction with a batch reactor would extract product at source, stabilize potentially labile products, and maximize system productivity by elimination of product feedback inhibitions. Adoption of a feed and bleed system could integrate a re-circulating fluidized bed loop with the continuous cell culture system discussed in Section 4.1, thereby eliminating the need for the sedimentation chamber. Preliminary work (33) has indicated that product adsorbtion is possible from biological suspensions by exploiting both ion exchange or immunoaffinity adsorbents in a fluidized bed. Washing of charged adsorbents is undertaken in fluidized mode, but efficient and concentrated desorption is best achieved in a fixed bed. This can be effected by collapsing the fluidized bed by flow reversal.

4.3 Aqueous two-phase fractionation

Aqueous two-phase systems, wherein mixtures of polymers (polyethylene glycol, dextran acrylamide, etc.) or polymers and salts above critical concentrations form two water-rich phases, have been recently applied as biocompatible liquid–liquid systems in protein processing (34). Conditions can be empirically established to predispose the partition of target proteins and impurities in the form of cell debris, nucleic acids, or protein contaminants into opposite phases. The molecular basis for this behaviour is not well understood, but descriptive models have been assembled which can exploit quantifiable characters of molecular mass, isoelectric point, and surface hydrophobicity in the establishment of working systems (35). The technique is claimed to scale well from laboratory to pilot plant with excellent recoveries. A major advantage lies in the potential for differential partition of cell debris (intractable in centrifugation) from soluble protein in simple liquid–liquid separators. As with liquid fluidized bed adsorbtion, applications beckon in the direct processing of whole broths or cell lysates rich in fine debris.

The high water content of phases, the low interfacial tension between phases, and the stabilizing effect of constituent polymers help to account for the high product recoveries. Degrees of purification are less impressive, and the technique best acts as a combination of primary separation and concentration steps which condition feedstocks for subsequent product fractionation. High salt and polymer content means that these steps should be chosen carefully. Consideration should also be given to methods of polymer recovery and recycle in larger-scale operations.

There is little in the literature to date demonstrating applications of the technique in conjunction with animal cell culture. One study (36) concluded that the molecular nature of products and impurities in serum-based culture of murine hybridomas for monoclonal antibody production was not particularly suited to efficient purification from clarified cell culture feedstocks. However, selective partition could be enhanced by dosing the phases with affinity chromatography matrices derivatized with antibodies specific for system components. This is at variance with the conventional approach of similarly derivatizing constituent polymers to enhance selectivity (37), but confers certain processing advantages upon the separation process (particle handling, ligand recovery, product concentration at elution, etc.).

4.4 Bioselective high performance liquid chromatography methods

Little previous mention has been made of high performance liquid chromatography (HPLC) in the recovery of proteins from animal cell culture. This technique exploits similar phenomena to those described for ion-exchange and bioselective adsorbent chromatography, but exploits more refined, small-diameter solid phases (silica- or polymer-based) and elevated working pressures (38). There are many commercial kits available targeted upon specialist laboratory purifications. However, the expense of equipment and materials, combined with a level of refinement not always compatible with the processing of macromolecules from crude feedstocks, confines the technique to analytical or small-scale preparative roles. This contrasts with the rapidly expanding use of large-scale (generally reversed phase) HPLC for small organic molecules in pharmaceutical manufacture.

Reports have been made (39, 40) of the application of miniaturized preparative bioselective HPLC systems (0.1 ml columns) for the rapid, near on-line quantitation of analytes in process feedstocks. Particles derivatized with specific antibodies or protein A have been used for the quantitation of IgG, respectively, in animal cell cultures and blood. Systems, which could be based upon any specific interaction between analytes and a biochemical ligand, were simply designed to guarantee quantitative adsorption and dedsorption over extended repetitive operations. Accuracy can rival conventional ELISA determinations in foreshortened timescales (39). Application of this technology for the rapid quantitation of products, impurities, and process antagonists will greatly enhance the monitoring and control of separation processes. For example, automatic sampling and bioselective HPLC analysis could monitor and control the integrated cell culture and product device discussed in Section 4.1. Rates of vessel dilution, residence times in the sedimentation chamber, loading times for immunoadsorbents, and controlled recovery of products could all be controlled on the basis of accurate and continuous monitoring of antibody product in the system.

5. Conclusions

Current technical developments of discrete bioreaction chambers, defined fermentation media, and the genetic expression of foreign protein products in mammalian cell cultures will yield little practical benefit unless equivalent advances in biochemical recovery guarantee the efficient manufacture of target products to purity specifications demanded of defined applications. Regardless of the physical design of bioreactors, all culture output will require varying degrees of solid–liquid separation, feedstock de-watering, and product fractionation. Individual polishing steps must be tailored to particular product types and applications. Considerable advantage can be gained by combining conventional unit operations associated with such procedures to effect the compression of recovery sequences to the benefit of overall product yield. Direct and early integration of bioselective processes such as adsorbtion in fixed or fluidized beds, or aqueous two-phase partition with mammalian cell culture output currently offer the best prospect for suitable sequence compression. Miniaturized bioselective HPLC processes offer opportunity for the ready establishment of pragmatic biogenesis suited to near on-line monitoring and control of integrated cell culture and product recoveries.

Acknowledgements

The author wishes to acknowledge the valuable comments of Jon Huddleston and co-workers in the Biochemical Recovery Group, School of Chemical Engineering, University of Birmingham.

References

1. Mizrahi, A. (1986). *Process Biochem.*, August, 108.
2. Spier, R. (1988). *Trends Biotechnol.* **6,** 2.
3. Harris, E. L. V. and Angal, S. (ed.) (1989). *Protein Purification Methods: A Practical Approach.* IRL press, Oxford.
4. Harris, E. L. V. and Angal, S. (ed.) (1989). *Protein Purification Applications: A Practical Approach.* IRL Press, Oxford.
5. Shacter, E. (1989). *Trends Biotechnol.* **7,** 248.
6. Handa-Corrigan, A. (1988). *Bio/Technology* **6,** 784.
7. Chase, H. A. (1984). *J. Chromatog.* **297,** 179.
8. Mohan, S. B., and Lyddiatt, A. (1991). *Cytotechnology* (In press.)
9. Bonnerjea, J., Oh, S., Hoare, M., and Dunnill, P. (1986). *Bio/Technology* **4,** 954.
10. Murkes, J. and Carlsson, C. G. (1988). *Crossflow Filtration.* John Wiley and Sons, Chichester.
11. Sansome-Smith, A. W., Huddleston, J. G., Young, T. W., and Lyddiatt, A. (1989). *Inst. Chem. Engng Symp. Series* **113,** 209.
12. Huddleston, J. G., Sansome-Smith, A. W., Wu, D-X., and Lyddiatt, A. (1989). In *Proceedings of the International Conference on Membrane Separation Processes*, p. 253. BHR Group, Cranfield, UK.
13. Christi, Y. and Moo-Young, M. (1986). *Enz. Microb. Technol.* **8,** 194.
14. Le, M. S. and Howell, J. A. (1985). In *Comprehensive Biotechnology*, Vol. 2 (ed. M. Moo-Young), p. 383. Pergamon Press, Oxford.
15. Chan, M. Y. Y., Hoare, M., and Dunnill, P. (1986). *Biotechnol. Bioengng* **28,** 387.
16. Stryker, M. H., Bertolini, M. J., and Hao, Y.-L. (1985). *Advan. biotechnol. Processes* **4,** 276.
17. Foster, P., Dunnill, P., and Lilly, M. D. (1976). *Biotechnol. Bioengng* **18,** 545.
18. Vijayalakshmi, M. A. (1989). *Trends Biotechnol.* **7,** 71.
19. Sassenfield, H. M. (1990). *Trends Biotechnol.* **8,** 88.
20. Sofer, G. K. and Nystrom, L. E. (1989). *Process Chromatography: A Practical Guide.* Academic Press, London.
21. Lambe, C. A. (1986). *Bioactive Microbial Products III—Downstream Processing* (ed. J. D. Sowell, P. J. Bailey, and D. J. Winstanley), p. 191. Academic Press, London.
22. Scopes, R. (1982). *Protein Purification: Principles and Practice.* Springer-Verlag, New York.
23. Chase, H. A. (1988). *Advan. Biotechnol. Processes* **8,** 163.
24. Dean, P. D. G., Johnson, W. S., and Middles, F. A. (ed.) (1985). *Affinity Chromatography: A Practical Approach.* IRL Press, Oxford.

25. Lyddiatt, A. (1990). In *Laboratory Methods in Immunology*, Vol. 2 (ed. H. Zola), p. 181. CRC Press, Boca Raton, Florida.
26. Desai, M. A. and Lyddiatt, A. (1990). *Bioseparation* **1,** 43.
27. Yarmush, M. L., Antonsen, K., and Yarmush, D. M. (1985). In *Comprehensive Biotechnology,* Vol. 2 (ed. M. Moo-Young), p. 489. Pergamon Press, Oxford.
28. Janson, J.-C. and Hedman, P. (1982). In *Advances in Biochemical Engineering* (ed. A. Fiechter), p. 43. Springer-Verlag, Heidelberg.
29. Lambert, K. J. (1989). *J. Chem. Technol. Biotechnol.* **45,** 45.
30. Lyddiatt, A., Desai, M. A., Huddleston, J. G., Rudge, J., and Shojaosadaty, S. A. (1989). *J. Chem. Technol. Biotechnol.* **45,** 47.
31. Wells, C. M., Patel, K., and Lyddiatt, A. (1987). In *Separations for Biotechnology* (ed. M. J. Verrall and M. J. Hudson), p. 217. Ellis-Horwood, Chichester.
32. Draeger, N. and Chase, H. A. (1990). *Inst. Chem. Engng Symp. Series* **118,** 129.
33. Wells, C. M. (1990). Unpublished DPhil thesis. University of Birmingham.
34. Kula, M.-R. (1985). In *Comprehensive Biotechnology,* Vol. 2 (ed. M. Moo-Young), p. 451. Pergamon Press, Oxford.
35. Huddleston, J. G., Ottomar, K. W., Ngonyani, D. M., and Lyddiatt, A. (1991). *Enzymes Microb. Technol.* (In press.)
36. Huddleston, J. G. and Lyddiatt, A. (1989). In *Separations Using Aqueous Two Phase Systems* (ed. D. Fisher and I. U. A. Sutherland), p. 309. Plenum Press, New York.
37. Ku, C-A., Henry, J. D. Jr, and Blair, J. B. V. (1989). *Biotechnol. Bioengng* **33,** 1081.
38. Kennedy, J. F., White, C. A., and Rivera, Z. (1988). *Int. Indust. Biotechnol.* **8,** 15.
39. Shojaosadaty, S. A. and Lyddiatt, A. (1987). In *Separations for Biotechnology* (ed. M. J. Verrall and M. J. Hudson), p. 436. Ellis-Horwood, Chichester.
40. Livingston, A. G. and Chase, H. A. (1989). *J. Chromatog.* **481,** 159.

10

Products from animal cells

BRYAN GRIFFITHS

1. Introduction

The manufacture of products from animal cells is currently a large and important component of the biotechnology industry. This exploitation of cells (*Table 1*) as a manufacturing substrate dates back to 1949 when Enders and his colleagues (1) published the first report on growing viruses in cultured cells. The effect of this discovery was dramatic, especially in the field of virology and eventually biotechnology. It took only 5 years for the first cell vaccine to be on human trial and licensed (Salk polio vaccine—1954). Classically, cell culture became the means of assaying, isolating, and producing a virus, and for 25 years this was the only cell product. The first human vaccines were produced in primary monkey kidney cells but this choice of substrate was controversial. Monkeys carry a wide range of adventitious viruses, many of them oncogenic (although subsequently it has been found not for human cells). The derivation of the human diploid cell (HDC) line WI-38 in the early 1960s by Hayflick and Moorhead (2) was welcomed as it made available a safer substrate for the production of human vaccines. This cell line was genetically stable, diploid, of normal cell derivation (fetal lung), free of all known viruses, and had useful markers for quality control evaluation. Unfortunately, it would only grow attached to a surface, and to low saturation cell densities. Thus manufacturers of vaccines had a 'safe' cell line to use, but a difficult one for the development of highly productive large-scale processes. However, as a result of it becoming licensed many new vaccines were rapidly developed during the 1960s. Meanwhile veterinary vaccines were being rapidly developed and, because of different licensing criteria, suspension cells such as baby hamster kidney (BHK) cells were being used as a substrate. This enabled production of foot-and-mouth disease virus vaccine (FMDV), produced in multi-thousand litre fermenters and the single largest cell product (2×10^9 doses annually). Human vaccines meanwhile were restricted to cells of normal derivation of either primary (monkey kidney, chick embryo cells) or human diploid origin. Culture systems were based on multiples of small units (flasks, roller bottles) or simple stirred suspensions of primary cells. The microcarrier culture system was developed in the 1960s by van Wezel to overcome this lack of scaleable unit process. Unfortunately, the development of this technique with reliable and efficient microcarriers took until the 1980s to perfect by which time manufacturers were 'locked' into the multiple systems by licensing regulations, or other more dramatic events had occurred which allowed alternatives to HDC to be used. Although the technique of deriving hybridoma cells for monoclonal antibody production was published in 1975 (3) the next major product from animal cells was in fact interferon. The significance of this product was that its perceived anti-tumour activity made large-scale production imperative in order to carry out clinical trials. Wellcome (UK) took the brave decision of developing lymphoblastoid interferon from suspension cells which could be grown in large bioreactors at that time being used for FMDV vaccine production at the 8000 litre scale. It was a brave step because the cell line was derived from a tumour (Burkitt's lymphoma); thus to get the interferon licensed would be a revolutionary step. This happened because of the efficiency of purification, the many (viral) inactivation steps, the ability to test the end-product and validate the process, and because of more realistic attitudes towards perceived risks on the part of regulatory authorities. This event broke the dogma that only human diploid (or normal primary) cells could be used for human biologicals.

The follow-up has been significant as subsequently a monoclonal antibody (OKT3) from a hybridoma (cancer) cell line, recombinant tissue-type plasminogen activator (tPA) from Chinese hamster ovary (CHO) (heteroploid) cells, and (limited licence only) polio and rabies vaccines from Vero (heteroploid) cells are all in clinical use. This is only the tip of the iceberg as over 100 other products (the majority monoclonal antibodies) are in clinical trial (*Table 1*).

In this chapter the range of animal cell products is reviewed with details of production and purification. The emphasis is on those in current use or clinical trial, but the selection is more wide-ranging than this to show the potential of animal cell biotechnology and the products of the future.

2. Product review

There are a vast range of products naturally expressed from animal cells *in vivo* including hormones and other regulatory molecules and metabolites. Some of these have been identified as important, or potentially important, therapeutic agents for reasons given below. Other cell products use the cell purely as a manufacturing unit, an example being viruses for vaccines. Also, of course, the cell itself can be considered a product, either for extraction of macromolecules or whole cells as artificial tissues (e.g. skin graft, pancreas).

Even cells specialized to produce a unique product express minute quantities, and most will not grow in culture, or express the product in culture. Thus the products in the most advanced stages of commercialization are those

Table 1. Development of animal cell culture as a production process

Year	Event
1949	Virus growth in cell culture (Enders, Weller, and Robins)
1954	Salk polio vaccine (monkey kidney cells)
1955	Sabin polio vaccine (monkey kidney cells)
1962	Establishment of human diploid cell line, WI-38 (Hayflick)
1963	Measles vaccine (chick embryo cells)
1964	Rabies vaccine (WI-38)
1967	Mumps vaccine (WI-38)
1969	Rubella vaccine (WI-38)
1970s	Experimental human vaccines (CMV, Varicella, TBE)
	Wide-range of veterinary vaccines
1975	Monoclonal antibody production (Kohler and Milstein)
1979	First recombinant cell line
1980	Scale-up to 8000 litres (IFN, FMDV vaccine)
1981	First monoclonal antibody diagnostic kit
1982	First recombinant pharmaceutical—insulin
1986	Lymphoblastoid γIFN licensed
	Recombinant animal cell products in clinical trial
	Polio and rabies vaccine from Vero cells
1987	OKT3 mAb licensed
1988	Recombinant tPA licensed
1989	Recombinant EPO in trial
1990	Recombinant products in clinical trial (HBsAG, factor VIII, HIVgp120, CD4. GM–CSF, EGF, mAbs, IL–2)

where there is either an available cell line with an artificially high secretion rate (e.g. cancer cells), or a means of artificially inducing higher production rates (e.g. interferons), or where the market drive was sufficient to give high priority to the development of recombinant cell lines for the product (e.g. tPA, erythropoietin (EPO)).

2.1 Viral vaccines

Between 1954 and 1970 a wide range of the familiar 'household' vaccines were developed and licensed. Initially, primary monkey kidney cells were used (for polio); however, concern over the adventitious simian viruses (oncogenic viruses) in these cells resulted in the human diploid cell line WI-38 becoming the dominant substrate. The viruses and cell lines used for human and veterinary vaccines are summarized in *Tables 2* and *3*.

Increased awareness of perceived dangers in vaccine preparations (oncogenes, retroviruses, DNA, transforming viruses, etc.) led to:

- increased quality assurance testing and costs
- greater difficulty in getting a new vaccine through clinical trials to a full product licence.

Table 2. Human viral vaccines

Vaccine	**Type**	**Date**	**Substrate**
Smallpox	L	1798	Animal skin
			Rabbit kidney cells
Rabies	I	1885	Mouse CNS tissue
		1956	Duck embryo
		1964	HDC
		1986	Vero
Yellow fever	I	1935	Chick embryo
Influenza	I	1936	Chick embryo
		Exp.	rCHO (HA subunit)
Polio	I	1954	Primary MKC
	L	1955	Primary MKC
	I	1984	Vero
Measles	L	1963	Chick embryo cells
Mumps	L	1967	HDC
Rubella	L	1969	HDC, rabbit kidney
Adenoviruses	I	1963	MKC, HDC(exptl.)
Cytomegalovirus	L	1974	HDC
Hepatitis B	I	1981	HBsAG from blood
		1986	rYeast
		1990	rCHO
Varicella	L	1974	HDC
T B Encephalitis	I	1975	Chick embryo cells
JE(B)	I	Exp.	BHK21
Rotaviruses	I	Exp.	MKC
Hepatitis A	I	Exp.	FRhK6, HDC
Herpes simplex	I	Exp.	HDC
Respiratory syncytial virus	I	Exp.	HDC
HIV	I	Exp.	rCells, anti-idiotype

L, live; I, inactivated; r, recombinant.

Table 3. Veterinary vaccines (selected examples from reference 24)

Vaccine	**Substrate**
FMDV	BHK cells; calf and pig kidney cells
Newcastle disease	Pig kidney; chick embryo cells
Marek's disease	Chick embryo
Rabies	Chick embryo and BHK cells
Pseudorabies	Pig kidney cells
Canine distemper	Dog kidney cells
Bovine diarrhoea	Embryonic kidney cells
Louping illness	Primary kidney cells
Bluetongue	Sheep kidney cells
Avian influenza	Vaccinia vector
Infect. bronchitis	Vaccinia vector

The punitive damage awards given by the courts to individuals suffering ill-effects from vaccines also deterred the development of new vaccines. However, the situation today has improved because:

- Purification technology has advanced.
- Biochemical assays of contaminants have increased in sensitivity.
- Current scientific information enables a more realistic evaluation of real and perceived dangers to be made.

Thus new vaccines are emerging either because alternative cell lines to HDC can be used (4), which makes the production process more economical (e.g. polio and rabies in vero cells) (compare *Protocols 1* and *2*), or due to the benefits of recombinant DNA technology (e.g. HIV, HSV). Recombinant vaccines will become increasingly available because of increased safety, increased immunogenicity, reduced side-effects, reduced production costs, safer production (without large quantities of pathogens), and their utility for viruses unable to grow in cultured cells. There are still a great many major human viral diseases without a vaccine (e.g. HIV, haemorrhagic fever, genital warts, genital herpes, respiratory syncytial virus, EBV–Burkitts lymphoma, nasal pharyngeal carcinoma, dengue fever). For a detailed description of viral vaccines, including the disease, diagnosis, production, efficacy and epidemiology, the reader is referred to the book by Plotkin and Mortimer (5).

Many of the common vaccines are produced by laborious multiple culture processes based on Roux flasks and roller bottles. This is because at the time of development and licensing there was no reliable large-scale unit process for human diploid cells. Roller bottles, although a labour-intensive method (*Protocol 1*), were reliable, easy to use with little to go wrong, and losses due to contamination were small. With proper organization of facilities, plants processing 28 000 bottles a week have been in operation for FMDV vaccine production (6).

Protocol 1. Production of human vaccines in human diploid cells using multiple culture scale-up

1. All processes start from a fully characterized working cell bank and the cells are expanded through a series of culture subdivisions until the production step is reached.

Population doubling[a]	*Polio vaccine*	*Rabies vaccine*
16	Cell seed bank (2×10^6/ml)	
17	Flask culture	
18	4 flask cultures	Cell seed bank
20	2 Roux cultures + 4 flasks	1 Roux culture

Protocol 1. *continued*

Population doubling[a]	*Polio vaccine*	*Rabies vaccine*
22	10 Roux cultures + 4 flasks	2 Roux cultures
24	42 Roux cultures + 4 flasks	6 Roux cultures
26	175 Roux cultures	24 Roux cultures
28		96 Roux cultures
29	85 Roller + 85 Roux cultures[a]	192 Roux cultures
30		384 Roux cultures
31		768 Roux cultures[b]

2. Cell growth was in the presence of medium. Now change to serum-free medium.
3. Infect with virus and allow the virus to replicate for 2–3 days.
4. Harvest virus. A non-lytic virus enables multiple harvests to be made.
5. Pool the harvested viruses.
6. Purify the virus where steps include filtration, concentration, inactivation, filtration, filling, freeze-drying, etc.

[a] These values represent the typical population doublings from the original isolation of the cells.
[b] The combined surface area of cultures is equivalent to 25 g microcarriers (10 litres).

Despite the availability of a reliable microcarrier system in the 1980s (see Chapter 1) few manufacturers have transferred to this technology. One reason for this is the need to re-license products manufactured by a different process, a very costly step to undertake. Some manufacturers, however, have taken up this challenge; for example, the Merrieux Institute in France, who initially produced FMDV vaccine on microcarriers (7), are now producing human polio and rabies vaccines using Vero cells (*Protocol 2*; reference 4).

Protocol 2. Preparation of viral vaccines—polio vaccine from Vero cells grown on microcarriers (Institute Merrieux, France; Reference 4)

Cell culture

The procedure is similar to that of *Protocol 1* except that unit volumes are increased rather than the number of cultures.

1. Start with a working cell bank (WCB) of 100×10^6 cells.
2. Proceed through 1, 5, 20, and 150 litres cultures to the final 1000 litre culture.
3. Inoculate the culture with virus.
4. Harvest the virus.

Purification

The aims of purification are to concentrate the virus and remove unwanted medium and cell components and, especially, viral DNA.

1. Perform the initial filtration (0.22 μm).
2. Concentrate the virus by ultrafiltration.
3. Perform an ion exchange (Spherodex).
4. Perform gel filtration (sepharose L-6B).
5. Repeat step **3**.
6. Inactivate the virus using formalin 1/4000.
7. Filter the virus.
8. Concentrate the virus.
9. Blend three monovalent vaccines.

Safety and Potency testing

Product testing is to prove safety, efficacy, and potency standards set up from the clinical trial data.

1. Check the identity by isoenzyme identification (LDH, PGD, GPI).
2. Check the karyology of the master cell bank only.
3. Test for tumourigenicity on Syrian hamster cheekpouch, nude mice, and new-born rats.
4. Test viral purity by testing for the absence of retrovirus.
5. Test for residual cell DNA using a radiolabelled DNA hybridization probe.
6. Test for potency using both monkey and human clinical trials.

2.2 Cytokines

The cytokines are a group of polypeptide molecules which regulate immune and inflammatory responses in the body. They include the lymphokines and the interferons (*Table 4*).

2.2.1 Lymphokines

Lymphokines are produced by leucocytes and act as mediators of cell immunity with hormonal functions which control many aspects of cell physiology. Many of the lymphokines are designated as interleukins which are involved in signalling between lymphocytes. When an antigen interacts with specific receptors on lymphocytes, interleukins and their receptors are expressed which direct the cells to proliferate, differentiate, and eliminate antigen, i.e. they

Table 4. Principal cytokines of therapeutic interest

Interleukins	
IL–1	Lymphocyte activating factor (LAF)
IL–2	T cell growth factor (TCGF)
IL–3	Colony stimulating factor (CSF)
IL–4	B cell stimulation factor 1 (BSF–1)
IL–5	T cell replacing factor (TRF)
IL–6	B cell stimulation factor 2 (BSF–2)
TNF	Tumour necrosis factor
G–CSF	Granulocyte colony stimulating factor
M–CSF	Macrophage colony stimulating factor
GM–CSF	Granulocyte–macrophage CSF
EPO	Erythropoietin
Interferons	
αIFN	(Alpha)-leucocyte, lymphoblastoid
βIFN	(Beta)-fibroblast
γIFN	(Gamma)-immune

are immune stimulants. Characteristic of all lymphokines are their high specific biological activity (1–100 pmol/litre) and correspondingly low levels in the circulation and body tissues. They are also capable of a rapid increase in circulating concentrations in response to specific signals.

The involvement of lymphokines with body homeostasis and dysfunction at the cellular level, particularly in haemopoietic regulation, makes them valuable potential therapeutic agents. The low level of expression from cells, and the fact that lymphocytes can only be grown/maintained in primary culture, have meant that investigative research has been limited by their availability and their use as drugs has been impossible. However, many of them have now been genetically inserted into cell lines which has benefited research into their structure and function. In fact, it has allowed the identification of about 16 lymphokines and has clarified the position whereby many of them have been known by a variety of names without the realization that they were identical (e.g. interleukin-1B, B cell activating factor, B cell differentiation factor, endogenous pyrogen, leucocyte endogenous mediator, serum amyloid A inducer, proteolysis inducing factor, catabolin, haematopoietin 1, mononuclear cell factor). Production from recombinant cell lines has meant a rapid development of many of them for clinical trial, mainly as anti-cancer agents (e.g. IL-2, G-CSF, GM-CSF). The one most advanced as a clinical drug is erythropoietin which is used on patients with anaemia, especially that due to renal disease. Erythropoietin is not a true lymphokine, being produced mainly by kidney cells as well as by macrophages. Its action is restricted to committed erythroid precursors only, which it stimulates to proliferate and form mature clones.

Production is by standard suspension (and microcarrier) techniques such as described for tPA in Section 2.4, although, being still in the developmental and clinical trial stages, the processes are highly confidential and very little information is available. One example is for IL-2 from a recombinant baby hamster kidney (BHK) cell line grown in serum-free medium (DMEM/F12 plus 5 mg/litre protein) on Cytodex-3 microcarriers (10 g/litre) in a continuous perfusion system (3 vol./day). The daily yield was 2.4 mg human IL-2 (8).

2.2.2 Interferons (IFN)

Interferons are a group of biologically active substances, initially of importance for their anti-viral properties, and subsequently for their immunoregulatory and anti-proliferative properties. They are of wide application in the treatment of viral infections and a range of cancers. There are three types of interferon (*Table 4*).

Alpha interferon

Sources of IFN-α are peripheral blood leucocytes, malignant leucocytes, and established lymphoblastoid cell lines, of which the Namalwa is the best known. As IFN-α is not glycosylated it is the most successful recombinant IFN produced in bacteria. However, because cells produce a collection of 8 to 11 subtypes and recombinant bacteria only one, the cell-derived material is more effective. Leucocyte IFN was the first to be produced in sufficient quantity for limited clinical evaluation (from buffy coats of virally induced leucocytes from human blood) (*Protocol 3*).

Protocol 3. Production of leucocyte (alpha) interferon

Interferon is produced from a primary culture of leucocytes (i.e. no cell growth is expected) which are induced to produce interferon by viral infection (analogous to *in vivo* responses).

1. Collect fresh human blood.
2. Remove buffy coats by centrifugation (1000 *g* for 10 min).
3. Store at 4°C overnight.
4. Re-suspend leucocytes in culture medium (1×10^7/ml).
5. Prime with IFN-α (125 units/ml) for 2 h.
6. Induce with Sendai virus (125 haemagglutinin units/ml) for 18 h.
7. Remove cells by filtration or centrifugation.
8. Adjust cell-free medium to pH 4 with HCl (inactivation).
9. Precipitate with 0.5 M KSCN, pH 3.5.
10. Re-suspend in acidified ethanol and centrifuge.

Protocol 3. *continued*

11. Repeat steps **9** and **10**. (Alternatively, if sufficient IFN present in sample, affinity chromatography based on controlled pore glass adsorption and elution with ethylene glycol and NK-2 sepharose chromatography can be used.)

Yield – 20 000 units/ml crude IFN.
– 13 000 units purified (approximate).

The large quantities needed to conduct full clinical trials, particularly against cancer, became available through the development of the Namalwa cell process. This B-type lymphoblastoid cell, derived from a patient with Burkitt's lymphoma, is easy to grow in stirred suspension culture and processes have been scaled-up to 8000 litres (*Protocol 4*; reference 9). The resultant IFN is identical to leucocyte IFN. In fact, this process represents a very important milestone in the development of animal cell technology as it was the first product to be licensed for human application from a tumour cell line. It broke the dogma that only normal (human diploid or primary) cells could be used as a substrate for human biologicals and has opened up the use of heteroploid cells for an increasing range of cell products, e.g. tPA, EPO, OKT3, mAb and vaccines. Thus, far more effective and cost-efficient cell processes can now be used based on suspension cells rather than low-density anchorage-dependent or primary cells.

Protocol 4. Production of lymphoblastoid (alpha) interferon

1. Start with a cell bank of Namalwa cells. (These cells grow well in culture and can be scaled-up to a large-volume production culture.)
2. Go through multiple culture passages in RPMI 1640/10% FCS using 100–8000 litre vessels.
3. Harvest cells.
4. Re-suspend in culture medium (1×10^6 cells/ml).
5. Prime with 2 mM sodium butyrate for 48 h (or 100 units/ml TFN).
6. Induce with Sendai virus (50 haemagglutinin units/ml).
7. After 18 h cool to 5–10°C.
8. Separate using continuous-flow centrifugation.
9. Precipitate with acid (TCA or HCl, 24 h).
10. Concentrate using ultrafiltration (20 000 mol. wt, pH 7).
11. Use gel filtration (Sephadex G75 or Ultrogel AcA-54) to remove high-molecular-weight components, including DNA.
12. Use affinity chromatography which allows one-step concentration and

purification of IFN by separating polyclonal and monoclonal (anti-human IFN on CNBr-activated sepharose) antibodies.

13. Stabilize with human serum albumin.

Beta interferon

Although many cells can be induced to produce IFN-β, human diploid cells have been the most used substrate (*Protocol 5*) mainly for reasons of safety. Production of IFN-β has been limited by the well documented short-comings of growing HDC in mass culture due to the fact that they are anchorage-dependent. Thus the usual array of multiple processes (e.g. roller culture, multitrays) have been used, as have large-scale unit processes (e.g. 300 litre stack plate and fixed bed bioreactors). Perhaps the most effective method is the microcarrier system which is currently operated by various companies at 50 litre (*Protocol 6*) and 4000 litre (10) scales. Production of IFN-β is shown in *Protocols 5* and *6*.

Protocol 5. Production of fibroblast (beta) interferon

Human fibroblasts are anchorage-dependent and are scaled-up using multiple cultures (flasks, roller bottles) as for vaccine production in *Protocol 1*.

1. Begin with a confluent layer of human diploid fibroblasts.

2. Prime with 100 units/ml IFN-β.

3. Induce:

(a) virally with Chikugunya virus (20–40 virus/cell) or Newcastle disease virus (20 haemagglutinin units/10^6 cells);

(b) non-virally using synthetic double-stranded RNA: Poly I:C (<100 μg/ml);

(c) using superinduction[a] with Poly I:C (5–20 μg/ml) for 2 h. To inhibit protein synthesis (of repressor) and allow IFN mRNA to accumulate, add cycloheximide (50 μg/ml) and leave for 5 h. Then remove cycloheximide. When it is removed, IFN is synthesized. To inhibit transcription and stabilize mRNA, now add actinomycin D (5 μg/ml).

4. Purify using a range of affinity techniques (11).

[a] Superinduction mimics viral infection using Poly I:C instead of virus. Poly I:C is an easily obtained chemical, i.e. no virus preparation or use is needed.

Protocol 6. Production of human β interferon on an industrial scale in microcarrier culture (10)

1. Begin with a master seed bank (population doubling (PDL) 18).

2. Expand to a working seed bank (PDL 25)—30×10^6 cells.

Protocol 6. *continued*

3. Expand using roller bottle expansion to PDL 32 (40 days).
4. Inoculate a spinner flask containing a suspension of microcarriers (4 g/litre) in culture medium (10 litre) and grow cells to PDL 33.
5. Subculture to two 50 litre bioreactors containing microcarriers and leave for 48 h.
6. Superinduce.
   (a) Wash cells with buffer.
   (b) Prime with human IFN (100 IU/ml) for 15 h.
   (c) Add Poly I:C (50 μg/ml) + cycloheximide (10 μg/ml).
   (d) After 4 h add actinomycin D (1 μg/ml).
7. Wash with buffer and add human albumin (0.025%); leave for 36 h.
8. Remove medium.
9. Place in a controlled pore glass (CPG) column and elute with step gradient acetic acid.[a]

[a] CPG chromatography is based on the hydrophobic properties of IFN-β adsorbing to glass at a neutral pH and then being eluted in acidic pH.

The complexity of the induction systems (especially in microcarriers where the poly I:C is absorbed on to many types) and inefficiency of culturing fibroblasts makes IFN-β a prime target for recombinant techniques. Unfortunately, neither bactrial nor yeast systems produce a particularly effective IFN-β; thus a recombinant mammalian cell system is the method of choice. Recombinant Chinese hamster ovary and mouse cell lines are available and processes are being developed to optimize product stability and fidelity. Incidentally, α and β interferon have different properties and are not alternatives.

Gamma interferon

IFN-α is obtained from T-lymphocytes, and also B-lymphocytes, as well as cell lines obtained from various leukaemias. However, yields are very low from these lines and production from human bloodleucocytes is the main source for large-scale production of IFN-α (*Protocol 7*).

Protocol 7. Production of gamma (immune) interferon (11)

1. Heparinize and centrifuge (1000 *g* for 10 min) fresh human blood.
2. Re-suspend cells in NH_4Cl (0.83%) at 4°C (30 min).
3. Centrifuge and re-suspend cells in medium + 20% FCS (3×10^5/ml).
4. Induce using PHA-P (50 μg/ml) or PMA at 37°C (4–7 days).[a]

5. Centrifuge culture (1000 *g* for 10 min); collect supernatant (expected yield 3–10 000 units/ml).
6. Concentrate, e.g. by ultrafiltration (10 000 mol. wt).
7. Purify using, for example:
   - CM–Sephadex C-50
   - gel filtration (Ultrogel AcA-54 or Sephacryl S-2000)
   - affinity chromatography (e.g. Con A–agarose)

[a] Phorbal esters (e.g. PMA) enhance α-IFN production.

2.3 Monoclonal antibodies

There are libraries devoted to the production of monoclonal antibodies (mAbs) thus a section of a chapter can only be highly selective. A more detailed description of the derivation of hybridoma clones, their mode of action, and their application is given in Chapter 6. Their production for research and commercial use in cell cultures will be the main topic of this section.

Initially, production of mAbs was in the ascitic fluid of mice which gives high yields (up to 10 mg/ml). However, apart from ethical issues, the use of large batches of animals becomes a laborious process for large-scale production. Thus *in vitro* culture methods have become the first-choice method. Hybridoma cells grow in free suspension; thus the scale-up methods available for such cells can be freely adopted. The largest (in volume) system currently in operation (2000 litres) is based on the airlift fermenter (at CellTech, UK; reference 12) and not in stirred bioreactors. This method was chosen because of its engineering simplicity, proportional scale-up kinetics, and low-shear/high-oxygenation characteristics. The yields from such a system are approximately 100 mg/litre (i.e. 50–100 fold lower than ascitic fluid). Many mAb producers have chosen the alternative strategy of using high-cell-density but low-volume cultures. There is a whole range of such cultures, e.g. hollow fibre reactors, ceramic cartridges (Opticell), membranes (Dynacell, MBR), porous macrocarriers (Verax), and encapsulation (see Reference 13 and Chapter 7 for a review of bioreactors). The similarity of these systems is that cell densities of 5×10^7 to 10^8/ml are achieved which means that mAb concentrations approaching that of ascitic fluid can be obtained. These processes operate in a continuous perfusion mode, often running for several months, and are thus capable of producing 1 g/day, i.e. 30–60 g per run. Although there is no saving in medium quantity, cheaper media can often be used because cells at high density are less dependent on serum/growth factors. A flow sheet for mAb production in a batch suspension system is summarized in *Protocol 8*.

Protocol 8. Production of mAbs in suspension culture

1. Start with cell seed ampoule in a T75 flask (RPMI 1640 + 10% FCS).
2. Scale-up to 10 litres in spinner flasks (e.g. 250 m → 1 litre → 2.5 litre → 5 litre → 10 litre working volume).
3. Inoculate fermenter (30–50 litres) at *ca.* 3×10^5/ml.
4. Grow cells to saturation density ($1–3 \times 10^6$/ml) over 3–4 days.
5. Maintain culture for 48–72 h for mAb production.
6. Remove cells by (continuous) centrifugation.
7. Concentrate supernatant (10–20×) by tangential flow filtration (100 000 mol. wt), e.g. Pellicon Unit.
8. Purify using protein A–sepharose affinity chromatography.
9. Elute IgG at pH 4.
10. Concentrate as in step **7**.

The principles are the same for whatever scale-up the final culture is, e.g. Celltech scale-up their airlift fermenters with a 10-fold increase per step (10 litres → 100 litres → 100 litres). An example of a small high-density system (Dynacell) is given in *Protocol 9*. This is a similar system to the hollow fibre bioreactor described in Chapter 7, Section 3. Monoclonal antibody culture has recently been reviewed (14).

Protocol 9. Small-scale perfusion type (Dynacell) production of mAbs

1. Assemble Dynacell culture according to manufacturers (Millipore) instructions.
2. Circulate RPMI 1640 medium through system to wet membranes and remove air bubbles.
3. Inoculate cell capsule (12.5 ml) with $1–3 \times 10^8$ cells/ml RPMI 1640 + 10% FCS.
4. Allow cell growth for 3 days, then perfuse with serum-free medium from reservoir (500–600 ml RPMI 1640).
5. Harvest medium every 3–4 days and replace with fresh medium (monitor by glucose limitation, harvest below 1 mg/ml).
6. Continue culture for 70+ days (150 days possible) until glucose utilization rate/mAb production rate falls significantly.
7. Yield: average production rate, 10 mg/day, 36 mg/litre. (Yield was *ca.* 25 mg/litre in flasks for this hybridoma.)

There have been considerable improvements in the purification methods for mAbs and this has been necessary to meet the therapeutic and affinity purification (of therapeutics) applications of mAbs. Pre-treatment is usually necessary to remove cell debris or, in the case of ascitic fluid, lipids. The efficiency of centrifugation is increased if flocculation of the cells is used. Concentration can be carried out by ultrafiltration (100-fold), or by ammonium sulphate precipitation (usually two cycles at 50% saturation for 30 min followed by centrifugation, 10 000 *g*). The pellet is re-suspended in PBS, dialysed against 10 mM Tris-HCl (pH 6.8), and then purified using ion exchange (anion or cation), affinity (protein A), or gel filtration chromatography (14). Affinity chromatography with Protein A (Protein A–sepharose CL-4B) is probably the most used technique (see *Protocol* 7, Chapter 5). This binds IgG-2a, 2b, and 3 strongly, but not IgG or IgM. Differential pH elution can separate these subspecies (eg. pH 4.5–5.0 for IgG2a; pH 3.5–4.0 for IgG2b; pH4.5 for IgG3). A Con A column can be used for IgM. A protocol for mAb purification using specific antigen–antibody affinity can be used in certain cases (see Section 4.1, Chapter 9). Ion exchange chromatography matrices include DEAE–5PW, Mono Q, and Accell QMA, and hydroxyapatite can also be used for purification by adsorption chromatography.

The importance of mAbs as a diagnostic tool, and then for purification technology has been the driving force for many new developments in cell biotechnology over the past 6 years—especially noticeable in the range of commercial culture systems available for producing mAbs. The need for mAbs will continue to increase, particularly to meet the demands of human therapy. One such product, an OKT3 antibody for kidney transplant rejection, is already licensed but there are over 100 more in clinical trial, particularly for cancer treatment. Advances in developing human myelomas, in humanizing mouse mAbs, and with recombinant DNA techniques for producing cell lines secreting human mAbs are such that the clinical applications will seem limitless. There are already many reports of human–mouse chimeric antibody being expressed from CHO cells and grown in 100 litre serum-free cultures. There is also the possibility of using mAbs as vaccines (anti-idiotypes), and as enzymes (e.g. esterases) (15). To meet this increasing production demand, culture systems will have to be scaled-up in volume (Celltech have plans for a 5000 litre airlift fermenter), and in unit density. Medium development has seen the batch/fed-batch cell concentrations increase from 1–2 $\times 10^6$/ml to 5–8 $\times 10^6$/ml. There are also now available culture systems capable of both supporting high cell densities (over 5 $\times 10^7$/ml) and operating at high volume (e.g. Verax, Cultispher, Siran, and Informatrix macroporous carriers; reference 16).

2.4 Plasminogen activators

Plasminogen activators catalyse the conversion of plasminogen to the active fibrinolytic enzyme, plasmin. This enzyme dissolves fibrin, the end-result of the coagulation cascade, to soluble fibrin degradation products (*Figure 1*). There is thus a critical balance and interdependence between the coagulation (see Section 2.5) and fibrinolytic cascades. Activators are widespread throughout the body and are associated with many physiological events, but their chief importance as a cell product is their role as a thrombolytic agent against occlusive vascular disorders (myocardial infarction, pulmonary and arterial embolism, etc.).

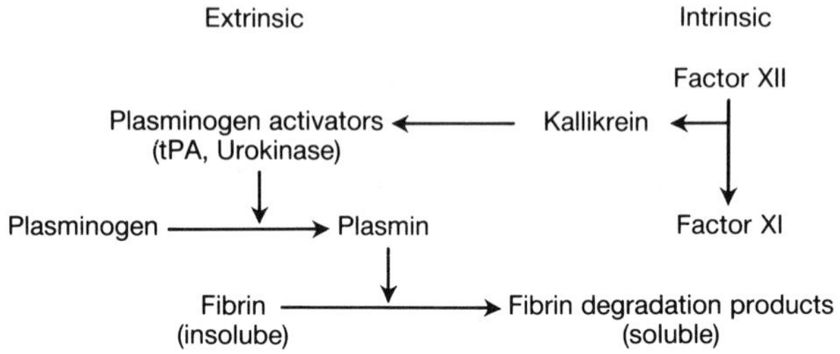

Figure 1. Part of the fibrinolytic cascade.

There are several types of plasminogen activators. First, there are the intrinsic factors present in the bloodstream, some of which are dependent on factor XII. Second, there is urokinase, an enzyme isolated from urine, which is relatively non-specific, activating both circulating and fibrin-bound plasmin indiscriminately, and which can cause unwanted haemorrhaging. Third, there are the tissue-type (tPA) activators secreted by a wide range of cell types, which are far more specific in their mode of action preferentially binding strongly to fibrin clots rather than circulating plasminogen. Mention should also be made of streptokinase, an extracellular protein released by some streptococcus strains, which has plasminogen activator properties and has been used therapeutically to a considerable extent. However, it is pyrogenic, antigenic, and, like urokinase, non-specific; thus alternatives have been actively sought.

Urokinase, extracted from urine, is available as a therapeutic agent but large-scale preparation is difficult. An alternative source is human primary embryonic kidney cells but, at a production rate of only 5–10 IU/10^6 cells/day, large-scale production is again difficult. A survey of large numbers of human cells showed that those derived from most normal adult tissues and tumours secreted urokinase, and those from melanomas and embryonic tissues secreted tPa. Many cell lines secrete both, e.g. human fibroblasts, leukaemic cells, and bovine endothelial cells (17). The discovery that the Bowes melanoma cell line secretes a tPA identical to that in normal human tissue, and at high levels (0.1 mg/litre) has enabled both investigative research into tPA and a large-scale culture process to be developed. Previously, the human uterus had been the prime source of tPA (0.01 mg/uterus, i.e. 10-fold less than 1 litre of Bowes melanoma cells).

Protocol 10. Production of tPA in concanavalin A (Con A) stimulated epithelial guinea-pig keratocyte (GPK) cells in perfused microcarrier culture (18)

1. Grow up cell seed in roller bottles (MEM + 10% FCS).
2. Transfer to microcarrier culture—12 g/litre cytodex 3 in 17.5 litre working volume fermenter with closed perfusion loop via a spin-filter and sparging 40% O_2 in air into cell-free compartment of spin-filter (13).
3. Inoculate at 2×10^4 cells/cm^2 in MEM + 10% FCS.
4. After 72 h when cell density is *ca.* 7×10^4/cm^2, replace medium with fresh serum-free MEM + Tween 80 (12.5 mg/ml) + Con A (50 μg/ml—optional).[a]
5. Grow for a further 60 h and harvest supernatant. Final cell density *ca.* 5×10^6/ml; tPA yield *ca.* 10 mg/litre.
6. Repeat steps **4** and **5** (if Con A present) and continue until cells are completely detached or in an advanced stage of necrosis (maximum three harvests).

[a] The action of concanavalin A in stimulating tPA is unknown, but it is active in the range 10 to 50 μg/ml. It not only enhances productivity but causes morphological changes. These are initially beneficial in decreasing cell detachment and prolonging the culture (to allow extra harvests to be made) but progressive necrosis occurs. Succinyl Con A is more efficient and less destructive on cell morphology and viability.

Although production protocols were developed for Bowes melanona tPA up to the 40-litre scale (*Protocol 10*), the problems of producing large enough quantities were still real. A dose of 7.5 mg was being used on patients and this needed 75 litres of culture. There was also the safety concern that the enzyme was derived from a tumour cell line. An alternative cell line of normal origin, epithelial GPK (17, 20, 21), was investigated. To overcome the lower yield always associated with normal, as opposed to tumour, cells various stimulatory agents were discovered, e.g. Con A that raised the expression level 10fold to that normally found in tumour cells. A production protocol based on a repeated-batch microcarrier culture is given in *Protocol 11*. Although many purification methods have been used for tPA (see review in reference 17); the most used technique is given in *Protocol 12*.

Protocol 11. Production of tPA from Bowes melanoma cells in continuous microcarrier culture (21)

1. Begin with a working cell bank.
2. Scale-up cell seed to 3×10^8 cells.
3. Inoculate 3 litre microcarrier culture (3–5 g/litre cytodex), pH 7.2.

Protocol 11. *continued*

4. Stir at 60–100 r.p.m. and perfuse via a spin-filter.
5. Subculture to 10 litres—add fresh medium and microcarriers to old medium + 10% serum.
6. Subculture to 40 litre microcarrier culture.
7. Grow cells to stationary phase (*ca.* 3×10^6 cells/ml) and change medium + 0.5% serum only.
8. After 24 h start perfusion (1 vol./day).
9. Continue until tPA yield drops (500–700 h).
10. Daily yield of tPA is 15 IU/ml (4 mg total).
11. Purify tPA by Sephadex G-50, zinc sepharose (4B) (elute with 0.05 M imidazole), Con A–agarose (1 M NaCl, 0.01% Tween 80), elute with a-D methylmannoside (0–0.04 M)/KSCN (0–2 M) gradient, dialyse against PEG (20 000 mol. wt), Sephadex G-150 in 0.01 M phosphate buffer + 1.6 M KSCN.

Protocol 12. Purification[a] of tPA from epithelial cells (21)

1. Clarify culture supernatant (centrifugation 500 *g* for 15 min).
2. Concentrate using ultrafiltration (Amicon 10 000 mol. wt hollow fibres)—5 × concentration.
3. Purify using zinc chelate agarose chromatography (20 mM Tris-HCl, pH 7.5; elution with 0.05 imidazole gradient).
4. Use concanavalin A agarose chromatography (0.01 M sodium phosphate, pH 7.5, 0.01% Tween 80. Elution with 2 M KSCN, 1 M a-D-methylmannoside).
5. Submit to dialysis (solid polyethylene glycol, MW 15–20 kd).
6. Initiate gel filtration (Ultrogel ACA-44, 0.01 M sodium phosphate, pH 7.5, 0.01% Tween 80, 1.6 M KSCN).
7. Repeat dialysis (0.15 M NaCl with 0.001% Tween 80).

[a] A protease inhibitor (e.g. aprotinin) is added to buffers to reduce proteolytic degradation of tPA, and Tween 80 is added to reduce loss by adsorption on to glass and plastic surfaces.

The only sensible way forward was to have a recombinant cell system. The gene from the Bowes melanoma line has been cloned into a number of cell lines (*Table 5*) with increased expression levels.

However, it was shown that doses of about 80 mg recombinant tPA were needed in patient trials. Currently, Genentech have a suspension production process at the 10 000-litre scale for the production of recombinant tPA

Table 5. Cell lines expressing tPA (19,20)

Cell type	Promoter	Amplification method	Peak titre (mg/litre)	Spec. product. (mg/10^9 cells/day
Recombinant				
YB2/3.9Ag20 (Rat myeloma)	RSV LTR	MTX/dhfr	52	4
CHO	SV40	MTX/dhfr	65	20
CHO	Ad2	MTX/dhfr	—	10
Mouse C127	BPV	Spontaneous	55	25
TRBM 6 (Human myeloma)	?	G418/neo	8	3
Bowes melanoma	SV40	MTX/dhfr	—	3
Native cells				
Bowes melanoma			1	0.3
GPK (epithelial)			1	0.4

(Activase), and which has a full international licence. However, there are many that feel the non-specific activators have many advantages, especially their speed of action upon inoculation to a patient suffering a heart attack, so that the issue is still not resolved as to what final product will get medical acceptance. The seriousness and widespread nature of occlusive diseases are such that the market is very big and over 50 pharmaceutical companies are competing for a share.

2.5 Blood clotting factors

Blood clots, or coagulates, following a complex cascade of protein interactions initiated by platelets at the site of the wound and culminating in the formation of a fibrin clot. Each step is one of amplification, i.e. much greater quantities of protein are produced at each subsequent step. The coagulation cascade is described in *Figure 2*, and any malfunction will prevent blood clotting and wound healing with a resultant heavy loss of blood. Problems arise in two of these steps through X chromosome linked defects. The first is a defect in factor IX, the other of factor VIII; thus, these two compounds are the principal factors to be produced for therapeutic use. Protein C, an anticoagulant, is also a targeted product.

2.5.1 Factor VIII

Factor VIII is a large glycoprotein complex which is normally present at very low concentrations (200 ng/ml). Components of this complex include protein VIII:C, which is a co-factor in the activation of factor X, and von Willebrand factor which is involved in platelet activation. Absence of factor VIII is the cause of haemophilia A, a serious haemorrhaging disease which leads to premature death unless treated.

Production of factor VIII has been by the Cohn fractionation process of plasma (22). This is based on treating plasma at 0°C with ethanol (8%) followed by changes in ethanol concentration and pH which separate out proteins (by precipitation) according to differences in solubility (see Chapter 2). Factor VIII is removed in the first fraction with fibrinogen. This process is amenable to large-scale production and has been a source of therapeutic factor VIII for 30 years. However, due to hepatitis B and then HIV contaminating the starting plasma, this method has become of reduced importance. Even with HIV screening, the risks of viral contamination are still there as each patient receives factor VIII (5–10 mg) from the equivalent of 600 litres of blood per year. The genes for factor VIII and von Willebrand have been isolated and cloned into animal cells (CHO, BHK21) with successful results. Factor VIII is a large molecule (mol. wt 265 000), which means that only animal cells could be considered as a host, but it is still an achievement to clone such a large gene. The annual dose per patient can be met by about 10 litres of culture. Von Willebrand factor, which is closely associated with factor VIII, has also been cloned in BHK cells.

2.5.2 Factor IX

Factor IX is a plasma glycoprotein (mol. wt 5700) which is part of the intrinsic clotting pathway where, in its activated form IXa, it interacts with factor VIII, phosopholipid, and calcium ions to form a complex that converts factor X to Xa (*Figure 2*). It is synthesized by liver hepatocytes where it undergoes various post-translational modifications before secretion into the bloodstream as a 415-amino acid protein. Because the post-translational modifications are complex and specialized, the belief is that Factor IX DNA clones would best succeed from a hepatic cell line. Thus it has been cloned into rat hepatoma (H4-11-E-C3), dog kidney (MDCK), hamster kidney, human hepatoma (Hep G2), as well as CHO cell lines. Expression levels are in the region of 300 ng/10^6 cells (where published). The maximum production rates in culture are from growing cells in serum-containing media, and yields are improved in controlled microcarrier culture compared to flasks. The clinical use of factor IX is to treat haemophilia B (Christmas disease), which is an X-linked bleeding disorder caused by a defect in clotting factor IX.

2.6 Hormones

There are a number of cell lines and primary cell cultures which express hormones (*Table 6*) but all in very low concentrations. There are also many tumour cell lines expressing ectopic hormones (i.e. hormones produced by cells not normally associated with hormone secretion; see *Table 6*). There are a number of hormones of distinct interest, for example, growth hormones in agriculture feeds, and, for humans, growth hormone (hGH), gonadotrophin, luteinizing hormone, follicle-stimulating hormone, and thyrotrophin. Insulin is the single largest hormone used clinically and is produced from animal

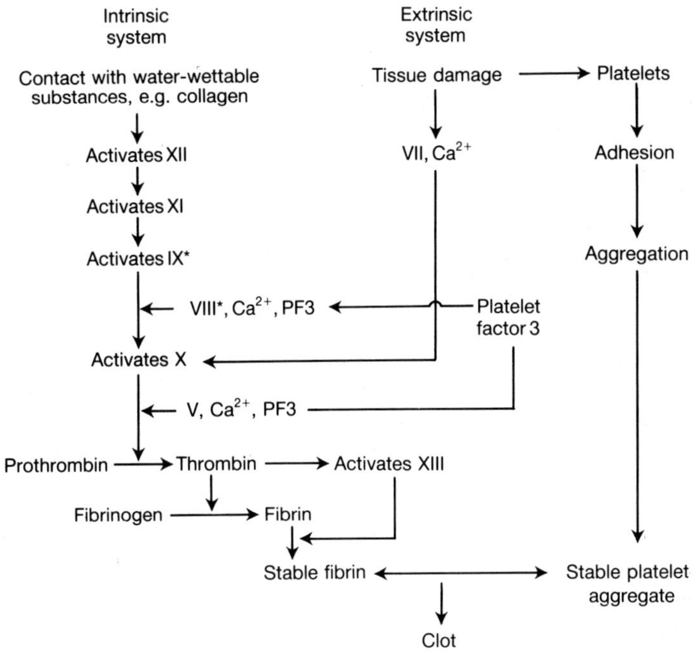

Figure 2. Blood coagulation cascade showing the roles of factors VIII and IX*.

Table 6. Cellular sources of hormones (23,24)

Hormone	**Cell source**	**Ectopic source**
Growth	Rat pituitary	
Prolactin	Rat, human pituitary	
Luteinizing (LH)	Human ant. pituitary	
Follicle stimulating (FSH)	Sheep pituitary	
Thyrotrophin (TSH)	Sheep pituitary	Hydatiform mole
Parathormone	Parathyroid adenoma	Kidney, bronchus
Insulin	Fetal rat pancreas	Kidney, bronchus
Gonadotrophin	Bronchus, pituitary, HeLa ovary, choriocarcinoma	Liver, bronchus
Adenocorticotrophin (ACTH)	Bronchial carcinoma, pituitary, HeLa	Bronchus, thymus
Corticotrophin		Bronchus, pancreas
ADH	Adrenal cell lines	Bronchus
Prostaglandins	Endothelium, kidney pituitary	
Glucosaminoglucans	Cornea, mastocytoma	
Calcitonin	Parathyroid tumour	

pancreas extraction (often with synthetic modification), or from recombinant prokaryotic systems—in fact it was the first recombinant product licensed (in 1982). The only hormone to be in advanced stages of production in recombinant animal cells is hGH (25).

2.6.1 Human growth hormone (hGH)

Human growth hormone has been prepared from human pituitary glands for the treatment of human dwarfism. Supplies were limited, pituitary glands being collected from cadavers in the mortuary. However, the danger of virus contamination, in particular Creutzfeldt–Jakob disease, has been the principal reason for seeking alternative soures. A recombinant hGH from bacteria was licensed in 1985; it is similar in structure to native hGH and appears to be equally efficacious. hGH is given as a weekly treatment to a patient for many years; thus, more attention to cumulative side-effects has to be given than in the case of a vaccine which has a very infrequent dosage. One problem may be the antigenic difference of the prokaryotic preparation. The protein, although large (mol. wt 20 000), is non-glycosylated so prokaryote expression gives a product with good biological activity. However, a gradual replacement with yeast and mammalian cell products is expected. A recently reported system (25) in CHO cells demonstrates the high expression level (200 mg/litre) that is possible from recombinant CHO cells; this expression is obtained in serum-free medium and a scale-up system using microcarriers. A C127 mouse cell transformed with bovine papillomavirus (BPV) vector is also being used to produce recombinant hGH.

2.7 Polypeptide growth factors

There are a wide range of polypeptide growth factors secreted by, and having a regulatory effect on, all cells within the body. They differ from hormones only by the fact that they are not synthesized in specialist endocrine organs. Most of them are of interest primarily in the elucidation of cell regulation to further the understanding of cancer, and as growth, or differentiation, factors in cell culture (26, 27). The ones of primary interest are listed in *Table 7*, of which platelet derived growth factor (PDGF), epidermal growth factor (EGF), and insulin-like growth factors (IGF) are the most investigated. EGF is valuable in the development of human epidermis in culture for subsequent transplantation. EFG is being developed as a drug to improve wound healing, especially in eye surgery, and is already in Phase III clinical trial. These growth factors would be more extensively used in culture as a serum replacement and to enhance productivity, especially in large-scale cultures, if they could be produced more cheaply. The hope is that this will happen as a result of recombinant DNA technology. The interaction of these growth factors and their receptors is very complex and the reader is referred to references 26 and 27.

Table 7. Polypeptide growth factors (GF)

EGF	Epidermal	Cell proliferation
FGF	Fibroblast	Mesodermal cell growth
NGF	Nerve	Nerve cell development
PDGF	Platelet-derived	Endothelial cell growth
IGF-I	Somatomedin C	Growth of cultured cells mediates growth hormone
IGF-2	Somatomedin A/MSA	Cell growth, differentiation
TGFα/β	Transforming	Suspension growth in agar, wound healing
ECGF	Endothelial cell	Endothelial cell maintenance mitogenic, tumour metastasis
	Transferrin	Iron binding and transport

2.8 Carcinomembryonic antigen (CEA)

Early diagnosis of cancer, particularly of those types which are easily operable, such as colon cancer, is an essential prerequisite for the control and cure of cancer (monitoring of therapy). Although there are many tumour markers they are tumour-associated, rather than tumour-specific, antigens. The most widely used marker for tumour diagnosis is the carcinoembryonic antigen (CEA), a high-molecular-weight glycoprotein secreted by adenocarcinoma cells. However, because false positives and negatives are obtained with CEA, it cannot be used as a cancer screening marker. Nevertheless, it is widely used to monitor therapy in patients with diagnosed cancers. Meanwhile, whilst the search continues for alternative markers, CEA production, and characterization, is being used as a prototype for the future manufacture of other cell-line-derived markers. This subject has been well reviewed (28), with details of CEA-producing cell lines, alternative production culture systems, and purification and characterization studies which have led to the raising of unique mAbs which give greater diagnostic reliabilty.

2.9 Cell as product

2.9.1 Screening anti-cancer drugs

Pharmokinetic studies on anti-cancer drugs added to proliferating tumour cells in culture have been successfully extrapolated to predict the response of the drugs in experimental animal models, and are a vital part of pre-clinical evaluation. The soft-agar tumour-colony-forming cell assay has been very successful. It permits the screening of drugs that can prevent metastasis. However, the technical problem of getting single-cell suspensions prepared from a patient's tumour biopsy is the factor limiting a more widespread use of this approach (e.g. many tumours do not contain clonogenic cells). Preclinical screening of anti-cancer drugs against a panel of human tumour cell lines and xenografts is still a valuable aid to the selection of suitable drugs.

2.9.2 Toxicity assays

Although many compounds have been tested in cell culture using cell numbers/viability/death/chromium^{51} release as the index, criticism can be levelled at the interpretation and validity of the results. Usually non-specialized cell lines are used with no allowance for the many interacting factors which occur *in vivo*. Thus, a move to more specialized cells with specific differentiated functions that can be quantified is the aim of all assay development studies. Hepatocytes are one of the most promising cell types for this purpose as they synthesize specific proteins, and the rate of synthesis is sensitive to toxic agents. Another possibility is to use recombinant cell lines with a specific function (e.g. cytochrome C) that is measurable and can be meaningfully extrapolated to the *in vivo* test.

2.9.3 Cell transplantation

When an organ, such as liver or pancreas, misfunctions or fails, a convenient therapy would be to add functional replacement cells. However, the problem has been to establish the cells *in vivo* without immune rejection. Several promising techniques to overcome this problem are now being assessed. Examples are: encapsulating the cells in alginate or agarose spheres which allows secreted products to diffuse through the sphere membrane whilst the cells are protected from the immunosurveillance system (e.g. artificial pancreas); to attach cells to collagen microcarriers (Cytodex 3) and transplant into the body cavities—this has been experimentally demonstrated with hepatocytes for renal failure in rats. Another more controversial example is the injection of human fetal brain cells into patients with Parkinson's disease.

The most established technique is replacement skin, especially for burns patients. Human epidermal keratinocytes obtained by biopsy are serially cultivated with irradiated fibroblasts. Eventually, a stratified squamous epithelium is produced and this is then passaged to generate a large quantity of epithelium. The epithelium can be detached from the substrate with Dispase II and grafted to burns patients. This technique has successfully operated in Italy since 1980 for patients with second- and third-degree burns (29).

2.9.4 Other applications

Cells have been one of the prime assay systems for viruses since 1949, and this application is still widespread in research, commercial, and hospital diagnostic laboratories. Cells also are a source of many research materials including organelles and, of course, mRNA and DNA for recombinant work. Finally, also at the research level, they provide many specialized proteins and carbohydrates (especially extracellular matrix materials). The Hep G2 cell line is a prime example of the range of products that can be produced by a cell—it secretes over 17 human plasma proteins including albumin, α-fetoprotein, transferrin, fibrinogen, etc. (30).

Baculovirus expression of recombinant proteins from insect cell lines is becoming an increasingly popular production system. A wide range of products have already been cloned (e.g. HIV components, pol, gag, gp120, gp160) with the perceived advantage that this method is biologically safer (no contaminating human viruses and DNA) and cheaper (simpler medium and lower incubation temperatures). Another possible application of baculovirus is its use as an insecticide (31).

For a complete review of cell products see reference 32.

3. The production process

3.1 Overview

In describing the various cell products in the previous section, and concentration on the actual culture process and purification, the complexity of the production process is perhaps understated. To put matters in perspective, the actual cost of manufacturing a product is a relatively small percentage of the total product cost. A schematic of the production process is given in *Figure 3*, with the flow diagram within the box representing the actual developed production process. It is the items outside this box which represent the greatest costs of a product: namely, the developmental work; the capital costs of the facility built to the required standards of pharmaceutical production; and quality control, clinical trials, and getting regulatory approval.

Developing media (usually serum-free with low protein content), derivation of the cell line (cloning, testing for stability and fidelity of product expression), and producing the master and working cell banks constitute an understandable and justifiable programme. Also recognizable are the costs of process development and optimization including bioreactor design and process control monitoring. What is underestimated, except by those involved, is the extensiveness of the testing carried out on the product and the constituents of the process, not only to initially license the product, but also carried out on every batch run. Documents for a batch production can be 300–500 pages long. It is also outside the scope of this chapter to describe the wide range of quality control tests that have to be carried out to prove the safety and efficacy of the final product (e.g. references 4 and 33).

In any description of animal cell products the above facts should be mentioned in order to put the production process in perspective. This is to emphasize that the development of any process should be an integrated effort, that all aspects should be kept as simple as possible, and the knock-on effects of each component on another recognized (e.g. medium formulation on purification). It also explains why new products have a long developmental time (5–10 years from laboratory to market place) and the high costs of new medicinal products.

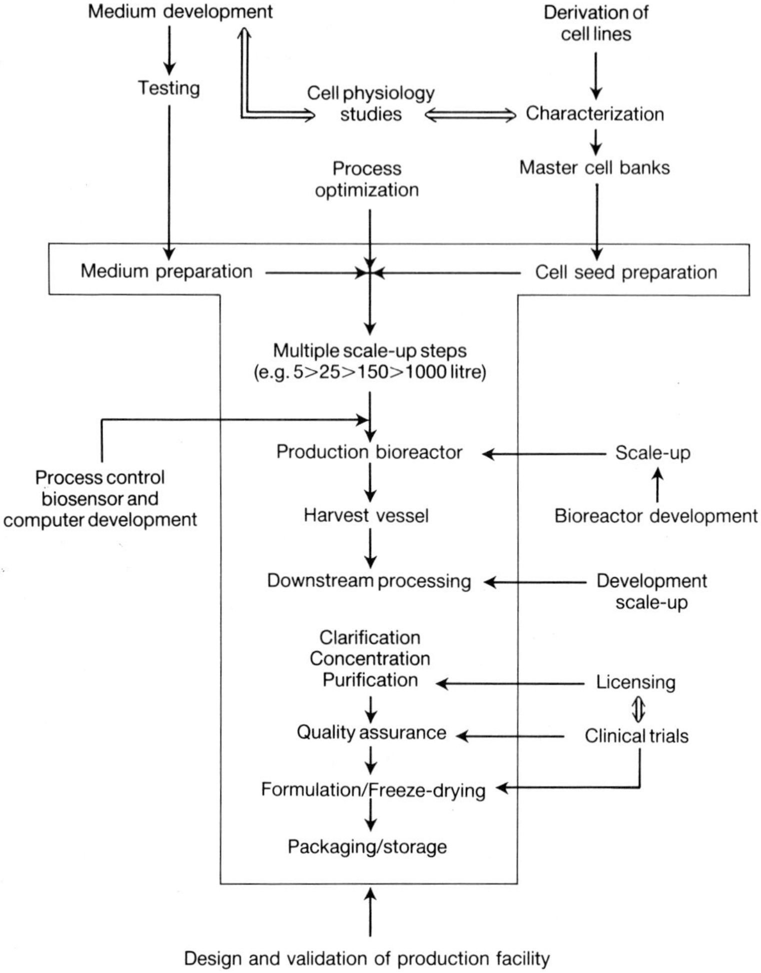

Figure 3. The production process—component interrelationships.

3.2 Choice of process

Two important issues which give rise to some debate in choosing a suitable culture process are: (a) high or low volume and cell density; and (b) batch or continuous process.

The highly successful large-scale processes currently in operation are mainly the large-volume/low-cell-density types, i.e. stirred bioreactors up to 10 000 litres (for IFN, tPA, FMDV vaccine) or airlift fermenters up to 2000 litres (for mAbs). These systems are relatively simple in engineering and process terms, and come with a huge background experience from the microbial industries. The move towards high-cell-density systems has been popular for mAb production because they need a smaller production facility and allow a quicker set-up time to get production started.

In order to grow and maintain cells at high densities an efficient perfusion system has to be used. A combination of perfusion and a complex, and expensive, immobilization strategy means that a continuously run process going on for many weeks, or months, is used. The advantages and disadvantages of batch versus continuous processes is a keenly argued and controversial issue. On balance, the unit cost of a product from a continuous process is 30–50% of that of a batch process. However, as pointed out in section 3.1, this may not be as significant as it seems. The disadvantages of continuous processes are that they are more complex and therefore more vulnerable to mechanical, or sterility, failure and compliance with licensing requirements, which is based on a batch system, is more difficult.

This controversy will become relatively unimportant because those high-density systems which cannot be volumetrically scaled-up will remain for specialized products not needed in kilogram quantities. To meet the expected demands for clinical products, process intensification will occur in methods capable of high-volume operation. This intensification will come about by improved medium formulations, by a better understanding of the genetic and physiological regulation of cells, and by the development of the porous microcarrier, or analogous technology. Porous microcarriers (such as Verax, Cultispher, Siran, Informatrix) allow cell density to be increased 20–100 fold over the solid microcarrier, and are equally suitable for suspension and anchorage-dependent cells (16). Solid microcarriers are currently used at scales up to 4000 litres so there is every possibility of scaling-up the porous microcarrier to 1000 litres—such a culture would support over 5×10^{13} cells and, with a productivity of 200 $\mu g/10^6$ cells, would produce over 2000 g of product. This product expression level, although excellent by current standards, is well below (by a factor of 5 or more) the theoretical expression levels of protein from cells obtained for short periods of time *in vivo* (34). Thus a combination of genetic manipulation, metabolic control by polypeptide growth factors, and a scaleable high-cell-density system may see production rates of 10 kg per day per 1000 litre reactor (35). This, I hope, is a glimpse of the future production process capability.

References

1. Enders, J. F., Weller, T. H., and Robbins, F. C. (1949). *Science* **109,** 85.
2. Hayflick, L. and Moorhead, P. S. (1961). *Exp. Cell Res.* **25,** 585.
3. Kohler, G. and Milstein, C. (1975). *Nature, London* **256,** 495.
4. Montagnon, B. J. (1989). *Dev. Biol. Stand.* **70,** 27.
5. Plotkin, S. A. and Mortimer, M. D. Jr (1988). *Vaccines* W. B. Saunders, Philadelphia.

6. Panina, G. F. (1985). In *Animal Cell Biotechnology*, Vol. 1 (ed. R. E. Spier and J. B. Griffiths), p. 211. Academic Press, Orlando, Florida.
7. Meignier, B., Meugeot, M., and Faure, H. (1980). *Dev. Biol. Stand.* **46,** 249.
8. Wagner, R., Krafft, H., and Lehmann, J. (1989). In *Advance in Animal Cell Biology and Technology for Bioprocesses* (ed. R. E. Spier, J. B. Griffiths, P. Crooy, and J. Stephenne), p. 374. Butterworths, Guildford.
9. Finter, N. B., Allen, G., Ball, G. D., Fantes, K. H., Johnston, M. D., and Lockyer, M. J. (1984). *Lab. Technol.* March/April, 157.
10. Morandi, M. and Valeri, A. (1988). *Adv. Biochem. Engng Biotechnol.* **37,** 57.
11. Mizrahi, A., Lazar, A., and Reuveny, S. (1990). In *Animal Cell Biotechnology*, Vol. 4 (ed. R. E. Spier and J. B. Griffiths), p. 413. Academic Press, London.
12. Birch, J. R., Thompson, P. W., Boraston, R., Oliver, S., and Lambert, K. (1987). In *Plant and Animal Cells—Process Possibilities* (ed. C. Webb and F. Mavituna), p. 162. Ellis Horwood, Chichester.
13. Griffiths, J. B. (1989). In *Animal Cell Biotechnology*, Vol. 3 (ed. R. E. Spier and J. B. Griffiths), p. 179. Academic Press, London.
14. Reuveny, S. and Lazar, A. (1989). *Adv. Biotechnol. Proc.* **11,** 45.
15. Carlsson, R. and Glad, C. (1989). *Biotechnology* **7,** 567.
16. Griffiths, J. B. (1990). In *Animal Cell Biotechnology*, Vol. 4 (ed. R. E. Spier and J. B. Griffiths), p. 147. Academic Press, London.
17. Griffiths, J. B. and Electricwala, A. (1987). *Adv. Biochem. Engng Biotechnol.* **34,** 147.
18. Kluft, C., van Wezel, A. L., van der Velden, C. A. M., Emeis, J. J., Verheijen, J. H., and Wijngaards, G. (1983). *Advan. Biotechnol. Proc.* **2,** 98.
19. Lubiniecki, A., Arathoon, R., Polastri, G., Thomas, J., Wiebe, M., Garnick, R., Jones, A., van Reis, R., and Builder, S. E. (1989). In *Advances in Animal Cell Biology and Technology for Bioprocesses* (ed. R. E. Spier, J. B. Griffiths, P. Crooy, and J. Stephenne), p. 442. Butterworths, Guildford.
20. Electricwala, A. and Griffiths, J. B. (1986). *Cell Biochem. Funct.* **4,** 55.
21. Griffiths, J. B., McEntee, I. D., Electricwala, A., Atkinson, A., Sutton, P. M., Naish, S., and Riley, P. A. (1985). *Dev. Biol. Stand.* **60,** 439.
22. MacLeod, A. J. (1988). *Adv. Biochem. Engng Biotechnol.* **37,** 41.
23. Katinger, H. W. D. and Bleim, R. (1983). *Adv. Biotech. Proc.* **2,** 62.
24. Griffiths, J. B. (1985). In *Animal Cell Biotechnology*, Vol. 2 (ed. R. E. Spier and J. B. Griffiths), p. 3. Academic Press, London.
25. Friedman, J. S., Cofer, C. L., Anderson, C. L., Kushner, J. A., Gray, P. P., Chapman, G. E., Stuart, M. C., Lazarus, L., Shine, J., and Kushner, P. J. (1989). *Biotechnology* **7,** 359.
26. *Cytotechnology* (1990). Volume 3(4). Kluwer Academic Press, Dordrecht.
27. Pimentel, E. (1987). *Hormones, Growth Factors and Oncogenes*. CRC Press, Boca Raton, Florida.
28. Epstein, N., Reuveny, S., Friedman, J., and Anel, N. (1990). In *Animal Cell Biotechnology*, Vol. 4 (ed. R. E. Spier and J. B. Griffiths), p. 445. Academic Press, London.
29. DeLuca, M., D'Anna, F., Bondanza, S., Franzi, A. T., and Canceda, R. J. (1988). *J. Cell. Biol.* **106,** 1919.
30. Knowles, B. B., Howe, C. C., and Aden, D. P. (1980). *Science* **209,** 497.
31. Miltenburger, H. G. and David, P. (1980). *Dev. Biol. Stand.* **46,** 183.

32. Butler, M. (1987). *Animal Cell Technology, Principles and Products*. Open University Press, Milton Keynes.
33. *Developments in Biological Standardization*, Vol. 70 (1989). Karger, Basel.
34. Murakami, H. (1990). *Cytotechnology* **3,** 3.
35. Griffiths, J. B. (1990). *Cytotechnology* **3,** 106.

A1

List of suppliers

Cell lines

American Type Culture Collection, 12301 Parklawn Drive, Rockville, MD 20852, USA.

European Collection of Animal Cell Cultures, PHLS, CAMR, Porton Down, Salisbury SP4 OJG, UK.

Hardware for cell culture

Amicon Ltd, Amicon House, 2 Kingsway, Woking, Surrey GU21 1UR, UK; 21 Hartwell Ave, Lexington, MA 02173, USA.

Coulter Electronics ltd, Northwell Drive, Luton, Bedfordshire LU3 3RH, UK; 590 W. 20 Street, Hialeah, FL 33010, USA.

Denley Instruments Ltd, Natts Lane, Billinghurst, Sussex RH14 9EY, UK.

Dynatech Labs, Daux Road, Billinghurst, Sussex RH14 9SJ, UK; 900 Slaters Lane, Alexandria, VA 22314, USA.

Flow Laboratories Ltd, PO Box 17, Second Avenue Industrial Estate, Irvine, Ayrshire, Scotland KA12 8NB, UK; 1710 Chapman Avenue, Rockville, MD 20852, USA.

Forma Scientific, PO Box 649, Marietta, OH 45750, USA

Heraeus Equipment Ltd, Unit 9, Wates Way, Brentwood, Essex CM15 9TB, UK; Postfach 1220, D-3360 Osterade am Harz, FRG.

Union Carbide UK Ltd, Cryogenics Division, Redworth Way, Aycliffe Industrial Estate, Aycliffe, County Durham, UK.

Watson-Marlowe Ltd, Falmouth, Cornwall TR11 4RU, UK.

Biochemicals/culture media

Amersham International, White Lion Road, Amersham, Buckinghamshire HP7 9LL, UK.

BDH Chemicals Ltd, Poole, Dorset BH12 4NN, UK.

Boehringer Mannheim GmbH, c/o BCL, Bell Lane, Lewes, East, Sussex BN7 1LG, UK; Biochemica, Postfach 310120, D-6800, Mannheim 31, FRG.

British Biotechnology Ltd, Watligton Road, Cowley, Oxford X4 5LY, UK; R & D Systems, 614 McKinley Place NE, Minneapolis, MN 55413, USA.

Flow Laboratories Ltd, PO Box 17, Second Avenue Industrial Estate, Ayrshire, Scotland KA12 8NB, UK; 1710 Chapman Avenue, Rockville, MD 20852, USA.

Gibco Laboratories, PO Box 35, Trident House, Renfrew Road, Paisley PA3 4EF, UK; 3175 Staley Road, Grand Island, NY 14072, USA.

ICN Biomedicals Inc, Free Press House, Castle Street, High Wycombe, Buckinghamshire HP13 6RN, UK; Biochemicals Division, PO Box 28050, Cleveland, OH 44128, USA.

Imperial Laboratories (Europe) Ltd, West Portway, Andover, Hampshire SP10 3L, UK.

Miles Scientific, 30W 475 North Aurora Road, Naperville, IL 60566, USA.

Northern Media Supply Company Ltd, Sainsbury Way, Hessle, N. Humberside HU13 9NX, UK.

Northumbria Biologicals Ltd, South Nelson Industrial Estate, Northumberland NE23 9HL, UK.

Sigma Chemical Company Ltd, Fancy Road, Poole, Dorset BH17 7NH, UK; PO Box 14508, St Louis, MO 63178, USA.

Cell culture consumables

Bio-rad Laboratories Ltd, Watford Business Park, Watford, Hertfordshire WD1 8RP, UK; 32nd & Griffin Ave., Richmond, CA 94804, USA.

Gelman Sciences, 600 S. Wagner Road, Ann Arbor, MI 48106, USA.

J. R. Scientific, PO Box 1862, Woodland, CA 95695, USA.

KC Biologicals, PO Box 14848, Lenexa, KS 66215, USA; 11 Chemin de Ronde, F-78110, Le Vesinet, France.

Lux Scientific Corp, 1157 Tourmaline Drive, Newbury Park, CA 91320, USA.

Millipore (UK) Ltd, The Boulevard, Blackmore Lane, Watford, Hertfordshire WD1 8YW, UK; The Millipore Corporation, Bedford, MA 01730, USA.

Müller-Lierheim, Biologische Laboratorien, Behringstrasse 6, 8033 Planegg/Munchen, FRG.

Percell Biolytica AB, Solvegatan 41, S-22370 Lund, Sweden.

Pfeifer & Langen, PO Box 100, 320—Frankenstr., 25 D-4047, Dormagen, FRG.

Pharmacia, Prince Regent Road, Hounslow, Middlesex TW3 1NE, UK; 800 Centennial Avenue, Piscataway, NJ 08854, USA.

Reactifs IBF, Societé Chimique Pointet-Girard, 35 avenue Jean-Jaures, F-92390, Villeneuve-La-Garenne, France.

Schott-Glaswerke, Hattenbergstrasse 10, Postfach 2480, 6500 Mainz, FRG.

Solohill Eng., 323 E. William, Suite 44, Ann Arbor, MI 48104, USA.

Sterilin Ltd, Sterilin House, Clockhouse Lane, Feltham, Middlesex TW14 8QS, UK.

Ventrex, 217 Read Street, Portland, ME 04103, USA.

Verax Corp, Etna Road, HC-61, Box 6, Lebanon, NH 03766, USA.

Fermenters/probes

Alfa Laval (Chemap), Great West Road, Brentford, Middlesex TW9 3DT, UK; 901 Hadley Road, South Plain Field, NJ 07080, USA.

Applikon Inc, 1165 Chess Drive, Suite G, Foster City, CA 94404, USA.

B. Braun, 13–14 Farmborough Close, Aylesbury, Buckinghamshire HP20 1DQ, UK; 999 Postal Road, Allentown, PA 18103, USA.

Bellco Glass Inc, 340 Edrudo Road, Vineland, NJ 08360, USA; c/o Horwell (Arnold R.) Ltd, 2 Grangeway, Kilburn High Road, London NW6 2BP, UK.

Dr W. Ingold AG, Industrie Nord, CH-8902 Zurich, Switzerland.

L. H. Fermentation, Porton House, Vanwall Road, Maidenhead, SL6 4UB, UK.

Life Sciences Laboratories, North Luton Industrial Estate, Sedgewick Road, Luton, Bedfordshire LU4 9DT, UK.

New Brunswick Scientific, 6 Colonial Way, Watford, Hertfordshire WD2 4PT, UK; 44 Talmadge Road, Edison, NJ 08817, USA.

Russell pH Ltd, Autermuchty, Fife, Scotland KY14 7DP, UK.

SGi (Setric), Allees de Bellefontaine, 15 F-31100, Toulouse, France; Newhaven BN9 OJX, UK.

Techne Ltd, Duxford, Cambridge CB2 4PZ UK; 3700 Brunswick Pike, Princeton, NJ 08540, USA.

Uniprobe Instruments Ltd, Clive Road, Cardiff, Wales CF5 1HG, UK.

Index

Index